Potential Use of Solar Energy and Emerging Technologies in Micro Irrigation

Potential Use of Solar Energy and Emerging Technologies in Micro Irrigation

Edited by:

Mulmudi Hemant Kumar

ARCLER
PRESS

www.arclerpress.com

Potential Use of Solar Energy and Emerging Technologies in Micro Irrigation

Mulmudi Hemant Kumar

Arcler Press
224 Shoreacres Road
Burlington, ON L7L 2H2
Canada
www.arclerpress.com
Email: orders@arclereducation.com

ISBN: 978-1-77407-704-7 (Hardcover)

ABOUT THE EDITOR

Hemant is the founder and director of BM Renewables Pvt Ltd based in Canberra. Dr. Mulmudi obtained his PhD in materials science (2014) from the Nanyang Technological University, Singapore, his MS in nanotechnology from the Ulm University, Germany (2009), and his bachelor's degree in metallurgical and materials engineering (2007) from the Indian Institute of Technology-Roorkee, India. He is an expert in the field of solar energy generation, especially, with the third generation of solar cells such as dye-sensitized solar cells and halide based perovskite solar cells. He worked on cathodoluminescence spectroscopy to unravel the optoelectronic properties of halide based perovskite materials. Dr. Mulmudi also developed cross-sectional nanoindentation technique to evaluate the fracture toughness of single and multi-junction photovoltaic devices at the Australian National University, Canberra. His research highlights include the first demonstration of a low-temperature, solution based processing of flexible solid state perovskite solar cells (Chemical Communications 49 (94), 11089-11091) and also stable lead-free inorganic perovskite solar cell (Advanced Materials 26 (41), 7122-7127). He has published over 35 internationally acclaimed peer-reviewed journals, filed a provisional patent with IP Australia (IP right number: 2018902843) and has co-authored a book titled 'Perovskite Solar cells: Technology and Practices'. He also delivers lectures on photovoltaic technologies as a casual academic at the Australian National University, Canberra.

TABLE OF CONTENTS

GLOSSARY

A

Agronomic – Agronomy is the science and technology of producing and using plants in agriculture for food, fuel, fiber, and land restoration.

AlternatingCurrent – Is an electric current which periodically reverses direction, in contrast to direct current which flows only in one direction.

Augmented – Having been made greater in size or value.

Avert – Turn away (one's eyes or thoughts).

B

Barren – (of land) Too poor to produce much or any vegetation.

Bubblers – a drinking fountain.

C

Canopy – is The aboveground portion of a plant community or crop, formed by the collection of individual plant crowns.

Conventional – Based on or in accordance with what is generally done or believed.

Cradle – A baby's bed or cot, typically one mounted on rockers.

Chemigation – Chemigation is the injection any chemical such as nitrogen, phosphorus or a pesticide into irrigation water and applied to the land using the irrigation system.

Cultivation – Is the science and art of cultivating plants and livestock.

D

Degradation – The condition or process of degrading or being degraded.

Democratization – The introduction of a democratic system or democratic principles.

Demonstration – An act of showing that something exists or is true by giving proof or evidence.

Desalination – Is a process that takes away mineral components from saline water.

Desertification – The process by which fertile land becomes desert, typically as a result of drought, deforestation, or inappropriate agriculture.

DirectCurrent – Is the unidirectional flow of an electric charge. An electrochemical cell is a prime example of DC power.

DripIrrigation – Drip irrigation is a type of micro-irrigation system that has the potential to save water and nutrients by allowing water to drip slowly to the roots of plants, either from above the soil surface or buried below the surface.

E

Emit – Produce and discharge (something, especially gas or radiation).

Enormous – Very large in size, quantity, or extent.

F

Foliage – Plant leaves collectively.

Fungicides – Are biocidal chemical compounds or biological organisms used to kill parasitic fungi or their spores.

Fertigation – Is the injection of fertilizers, used for soil amendments, water amendments and other water – soluble products into an irrigation system.

G

Grave – A hole dug in the ground to receive a coffin or dead body, typically marked by a stone or mound.

GreenRevolution – The Green Revolution, or Third Agricultural Revolution, is a set of research technology transfer initiatives occurring between 1950 and the late 1960s, that increased agricultural production worldwide, particularly in the developing world, beginning most markedly in the late 1960s.

Greenhouses – Also called a glasshouse, or, if with sufficient heating, a hothouse) is a structure with walls and roof made chiefly of transparent material, such as glass, in which plants requiring regulated climatic conditions are grown.

Grid – Electrical grid, electric grid or power grid, is an interconnected network for delivering electricity from producers to consumers.

GroundwaterManagement – Management of groundwater sub-basins to provide for multiple long-term benefits without resulting in or aggravating conditions that cause significant economic, social, or environmental impacts, such as long-term overdraft, land subsidence, ecosystem.

H

Horsepower – is a unit of measurement of power, or the rate at which work is done, usually in reference to the output of engines or motors0

Hose – is a flexible hollow tube designed to carry fluids from one location to another.

I

Immersed – Dip or submerge in a liquid.

Irrigation – Is the process of applying controlled amounts of water to plants at needed intervals.

L

Longevity – Is sometimes used as a synonym for "life expectancy" in demography.

M

MicroSprinklers – Is a method of providing rainfall-like irrigation to the crops.

Monocrystalline – Monocrystalline silicon, more often called single – crystal silicon, in short mono c-Si or mono-Si, is the base material for silicon-based discrete components and integrated circuits used in virtually all modern electronic equipment.

N

NetMetering – Net metering is a billing mechanism that credits solar energy system owners for the electricity they add to the grid.

P

PeakWattage – This value specifies the output power achieved by a Solar module under full solar radiation (under set Standard Test Conditions).

Photon – Is a type of elementary particle. It is the quantum of the electromagnetic field including electromagnetic radiation such as light and radio waves, and the force carrier for the electromagnetic force.

Photovoltaic – Is the conversion of light into electricity using semiconducting materials that exhibit the photovoltaic effect, a phenomenon studied in physics, photochemistry, and electrochemistry

Polycrystalline – Polycrystalline silicon, or multi crystalline silicon, also called polysilicon or poly-Si, is a high purity, polycrystalline form of silicon, used as a raw material by the solar photovoltaic and electronics industry.

Polymers – Is a large molecule, or macromolecule, composed of many repeated subunits.

R

RankineCycleEngine – Is a model used to predict the performance of steam turbine systems.

Reassembled-(of a group) Gather together again.

S

SemiaridClimate – Is the climate of a region that receives precipitation below potential evapotranspiration, but not as low as a desert climate.

Semiconductors – A semiconductor material has an electrical conductivity value falling between that of a conductor, such as metallic copper, and an insulator, such as glass.

SoilContamination – Soil contamination or soil pollution as part of land degradation is caused by the presence of xenobiotics chemicals or other alteration in the natural soil environment.

SolaTube – Is a day lighting system that are engineered to efficiently capture the sun's rays and bring that natural light into the house.

SolarPathfinder – Solar Pathfinder is the professional PV installer's tool for determining the most economical and efficient photovoltaic array location and position.

SolarPonds – A solar pond is a pool of saltwater which collects and stores solar thermal energy.

Sprinkler – An irrigation sprinkler is a device used to irrigate agricultural crops, lawns, landscapes, golf courses, and other areas.

T

Thermostats – Is a component which senses the temperature of a physical system and performs actions so that the system's temperature is maintained near a desired set point.

Topography – S the study of the shape and features of land surfaces. The topography of an area could refer to the surface shapes and features themselves, or a description.

Trivial – Of little value or importance.

V

VariableFrequencyDrive – Is a type of adjustable-speed drive used in electro-mechanical drive systems to control AC motor speed and torque by varying motor input frequency and voltage.

W

WaterConservation – Water conservation includes all the policies, strategies, and activities to sustainably manage the natural resource of fresh water, to protect the hydrosphere, and to meet the current and future human demand.

WaterManagement – Water resource management is the activity of planning, developing, distributing, and managing the optimum use of water resources. It is a sub-set of water cycle management.

LIST OF FIGURES

LIST OF TABLES

LIST OF ABBREVIATIONS

AC	alternating current
AKRSP	aga khan rural support programme
ASAE	American society of agricultural engineers
B	boron
BIA	Bureau of Indian Standards
Ca	calcium
CAB	commonwealth agricultural bureau
CAGR	compound annual growth rate
CO_2	carbon dioxide
CSP	concentrating solar power
Cu	copper
CWC	central warehousing corporation
DC	direct current
DOE	department of energy
EV	electric vehicle
FAO	Food and Agriculture Organization
Fe	iron
FHA	forced hot air
FICCI	Federation of Indian Chambers of Commerce and Industry
FMI	flood method of irrigation
GDP	gross domestic product
GHG	greenhouse gas
GIZ	GesellschaftfürInternationaleZusammenarbeit
GW	gigawatts
HDPE	high-density polyethylene
HP	horsepower

HVAC	heating ventilation and air conditioning
IAI	Indian Agricultural Institute
ICID	international commission on irrigation and drainage
IDE	international development enterprises
IDT	innovation diffusion theory
IEA	International Energy Agency
IEC	international electro-technical commission's
IoT	internet of things
IR	infrared radiations
ISO	International Organization for Standardization
IWMI	International Water Management Institute
K	potassium
LCD	liquid crystal display
Mg	magnesium
MI	micro-irrigation
MM	motivational model
Mn	manganese
Mo	molybdenum
MPCU	model of pc utilization
N	nitrogen
NGO	non-profitable government organizations
NMMI	national mission on micro-irrigation
OECD	Organization for Economic Co-operation and Development
Op-Amp IC	operational amplifier IC
P	phosphorus
PAEGC	powering agriculture: An energy grand challenge for development
PDS	pipe distribution irrigation system
PE	polyethylene
PLS	partial least squares
PMKSY	Pradhan Manti Krishi Sinchayee Yojana
PV	photovoltaic
PVC	poly-vinyl chloride
PVWPS	photovoltaic water pumping system
S	sulfur

SCT	social cognitive theory
SDI	subsurface drip irrigation
SIM	sprinkler irrigation method
SPIS	solar powered irrigation systems
SPMI	solar powered micro-irrigation
SPV	solar photovoltaic
SPVPS	solar photovoltaic pumping system
SPVWPS	solar PVWPS
TAM	technology acceptance model
TPB	theory of planned behavior
TRA	theory of reasoned action
UTAUT	unified theory of acceptance and use of technology
UV	ultra-violet
VFD	variable frequency drive
WaPOR	water productivity open-access portal
Wp	peak wattage
WRO	World Relief Organization
WUAs	Water Users Associations
WUE	water use efficiency
Zn	zinc

PREFACE

Agriculture is a huge contributor to the food security of the world and also the economy of several countries across the globe. Agriculture assumes prime importance being the source of one of the basic needs of the world.

In the modern world there is enormous pressure on the resources due to their gigantic usage across several forms. Thus, there is a huge pressure on the water resources as well. Water is one of the main components of agriculture and prospects of agriculture seem impossible without it.

Hence, keeping the low amounts of water available for the irrigation and agricultural purposes, the concept of micro-irrigation was developed. This kind of irrigation technique promised efficient usage of water resources and ensured that the irrigation of the agricultural lands was done in a sustainable manner.

However, to employ micro-irrigation technique various other sources of energy were needed. For sustainable use and for the optimized use of resources, alternative sources of energies were explored. Solar energy was deemed as a crucial source of energy in this sector along with some other similar renewable sources of energy.

This book makes the readers aware about the different kinds of sources of energy that can be employed in the various technologies concerning micro-irrigation. The book starts off by introducing micro-irrigation system to the readers so that they are well-aware about the subject that is being stressed on in the entire book.

The second chapter gives a review of solar energy to the readers to make them comfortable with the potential of this kind of energy and its various applications. The book then elaborates on the various kinds of benefits of irrigation that is powered by solar energy and the risks associated with such a technology.

The fourth chapter throws light on the subject of micro-irrigation management and talks about it in detail. Then next chapter takes the readers through the various kinds of technological advances in the field of micro-irrigation and informs about their application in the field.

The sixth chapter sheds light on the potential of solar energy to support micro-irrigation facilities by talking about the benefits of having such a resource and talking about green technology. The seventh chapter takes the readers through the sustainability aspect of employing solar power in the micro-irrigation facilities and emphasizes on its role to act as a good form of future energy.

The last chapter talks about the various kinds of innovations and the challenges that have been made in the field of micro-irrigation, so that the readers are well-aware of the latest developments and remaining hurdles in this subject.

With this book, I intend to educate the readers about the enormous possibilities of micro-irrigation and to update them about the role of solar energy in sustaining such a great prospect for agriculture. I hope that this book provides detailed insights and that it can provide the readers with the answers that they may be looking for.

Chapter 1

Introduction to Micro-Irrigation System

CONTENTS

This chapter explains about the micro-irrigation system. This system is defined as the modern system that is used for the purpose of irrigation. There are various classifications of irrigation systems. All the classifications of the micro-irrigation system are explained in this chapter. Also, all the features, the advantages and the shortcomings of the micro-irrigation system are properly explained.

Drip irrigation, which is a classification of micro-irrigation, is also further classified into various categories such as sub-surface drip irrigation, online drip, family drip, surface drip irrigation, in-line drip. All these categories are also explained briefly. Furthermore, some other kinds of irrigation systems like bubbler irrigation system and spray irrigation system are explained.

1.1. INTRODUCTION

Water is considered as one of the most important inputs for the field of agriculture. It consumes more than 80% of the water resources of the whole country. Thus, the availability of sufficient quantity as well as quality of water is the most important key factor for attaining higher levels of productivity (Figure 1.1).

Figure 1.1. Micro-irrigation is the modern method of irrigation.

Source: Image by Wikipedia.

There are some of the imperatives that a country needs to improve such as investments in conservation of water, enhanced techniques to ensure the timely supply, and enhance its efficient use. On the other hand, the poor irrigation efficiency of conventional irrigation system has not just decreased the anticipated result of investments that is made towards water resource development, but it also resulted in various environmental problems such as water logging and soil salinity.

In this way, it also affects the crop yields. Thus, there is the need for massive investments in adoption of enhanced techniques of irrigation like drip and sprinkler, which includes fertigation.

There are several options that are available for decreasing the demand of water in agriculture. Firstly, the supply-side management practices consist of watershed development as well as water resource development through medium irrigation projects, major projects, and minor irrigation projects.

The second one is through the demand management practices which consists of enhanced water management technologies or advanced practices. The micro-irrigation (MI) technologies like drip technology and sprinkler technology are the important interventions in saving the water and enhancing the productivity of the crop.

There is some evidence that shows that more than 40% to 80% of water can be collected and water use efficiency (WUE) can be improved up to 100% in MI system that is properly designed and managed as compared to 30–40% under conventional practice (INCID 1994; Sivanappan 1994; Kumar, 2012).

The term "micro-irrigation" explains about a family of irrigation systems that use water through small devices. All these devices deliver water onto the surface of the soil extremely near the plant or under the surface of the soil directly into the plant root zone.

Micro-irrigation is defined as a method or technique for delivering slow, frequent applications of water to the soil by the use of a low-pressure distributing system as well as a special flow-control outlet. In addition to it, micro-irrigation is also referred to as subsurface, drip irrigation, bubbler or trickle irrigation. It has to be noted that all these have same design as well as management criteria.

The systems help in delivering water to individual plants or different rows of plants. The outlets are kept at short intervals all along the small tubing and it is only the soil near the plant that is watered. The outlets consist of orifices, bubblers, emitters, and sprays or micro sprinklers with discharge that is ranging from 2 to over $200 lh^{-1}$.

It has to be noted that drip irrigation was developed initially as a sub-irrigation system and this fundamental concept underlying drip irrigation can be traced back to experimentations in Germany in the period of 1860's. The initial work in drip irrigation in the United States of America was a study carried out by House in Colorado in the year 1913. The essential breakthrough was made in Germany way back in the year of 1920 when there was introduced perforated pipe drip irrigation.

During the initial period of 1940's, an engineer from Israel, Symcha Blass discovered that a big tree near a leaking tap showed more vigorous growth as compared to the other trees in the region. After this discovery, he thought about the idea of an irrigation system that would help in using water in small quantity literally drop by drop.

It has been observed that the earliest drip irrigation system included plastic capillary tubes. These tubes were of small diameter (1mm) that are attached to large pipes. Coiled emitter was One of the refinements that was made by Blass in his original system. In the early period of the 1960's, the experiments that were conducted in the Israel reported spectacular outcome when Blass system was applied in the desert area of the Negev as well as Arava.

Micro-irrigation units are being installed widely across U.S.A, Australia, Israel, Mexico in their diverse forms. On the other hand, countries such as Canada, Cyprus, France, Iran, New Zealand, UK, Greece, and India are trying to integrate and adapt to the modern methods of micro-irrigation.

With the increase in the availability of plastic pipes as well as development of emitters in Israel, it has since become an essential technique or method of irrigation in Europe, Japan, Australia, Israel, Mexico, South Africa and the United States (INCID, 1994).

For example, it has been observed that in many developing countries such as in India, drip irrigation was practiced by the use of indigenous methods like perforated bamboo pipes, earthenware pipes and pitcher/ porous cups.

On the other hand, in the eastern region of India, some of the tribal farmers are making use of bamboo drip irrigation system for pepper, betel, and other nut crops by diverting hill streams in hill slopes. In places such as Rajasthan and Haryana, earthenware pitchers and porous cups have been used for the purpose of growing vegetable crops.

In many of the developing countries such as India, drip irrigation was introduced in the early 70's at agricultural colleges and some other research organizations. The expansion of drip irrigation has really achieved momentum in the previous decade.

These advancements have taken place particularly in areas of severe water scarcity and in commercial crops or horticultural crops, like banana, grapes, coconut, fruit trees, sugarcane, and plantation crops in some regions of Southeast Asia such as India.

It has been observed that micro-irrigation has been accepted particularly in the arid regions for the purpose of watering high value crops like grapes and other vine crops, fruits, and orchard trees, sugarcane, flowers, pineapples, strawberries, and vegetables.

It has also been observed that producers, growers, and landscapers have adapted micro-irrigation systems to suit their requirements for precision water application. Micro-irrigation systems are very popular not just in arid regions and urban settings but they are also popular in sub-humid as well as humid zones. In these sub-humid and humid zones, the supplies of water are limited, or it can be said that water is expensive. Micro-irrigation is widely used with ornamental plantings in urban landscapes.

In order to bring more area under irrigation, it has become vital to bring in new irrigation methods or techniques viz. Micro-Irrigation and Sprinkler Irrigation for saving the usage of water and increasing the productivity per unit of water.

In addition to it, this technology also arrests water logging as well as secondary salinization issues of the canal command areas and it also keeps the check on the receding water table and deteriorating quality of the water in the well command areas.

Micro-irrigation is to be considered as a complete plant support system that starts with planting material to post harvest management and then marketing. Thus, the need of micro-irrigation should be encouraged or promoted in a holistic manner which consists of good agronomic practices, correct cultivars, post-harvest handling, processing as well as marketing.

This leads to an end-to-end approach. It has to be noted that water source development as well as recharge of wells via watershed management would also form a part of the technology package (AGRICOOP, 2005).

1.2. OVERVIEW OF MICRO-IRRIGATION TECHNOLOGY

Water is carried to the actual root of the plant in micro-irrigation system. Water is not just carried to the surrounding dirt. It makes use of tubes, pipes, and a dripper in order to deliver the water slowly. In this method, much less amount of water as compared to other normal irrigation. Moreover, it is more efficient and even it is ecological.

It has been observed that in the earlier times, a clay pot which is having holes were filled with water and this pot was buried in the ground. After this, a clay pipe was used. In the end, it was changed into the more common perforated plastic tubing.

Even the latest advancements or developments consists of a plastic water emitter that is placed at the bottom of the root. This plastic water emitter helps in dripping out the water into the root. It has been observed that latest techniques or methods make this a valuable contribution to the world of agriculture, particularly those areas that are having the shortage of water.

Some other kinds of micro-irrigation consist of the bubbler and micro sprinkler. In the bubbler system, the drip is more spread out. On the other hand, the micro sprinkler is used overhead where the water is released in as micro-sprays. This is generally in a closed structure like a greenhouse.

It has to be noted that the irrigation tube is placed deep in the dirt, that is close to the surface. The emitter is placed in the tubing (which is perforated) at each plant.

Water is pressurized slowly by the pump through the emitters. If in case the drip irrigation system is combined with mulching, this type of watering would actually decrease the surface evaporation and it will also be quite efficient in conservation methods. In addition to it, drip irrigation can also help in reducing the foliage diseases. This foliage diseases come about with wet as well as moldy leaves because the water directly goes down to the main root of the plant.

In farms as well as in commercial greenhouses, micro-irrigation is used. It has proofed to be successful in a commercial sense because of the automation. Moreover, fertilizer can be added to the water with the help of piping as well as pressurized pumps.

This automatically sets the watering as well as feeding of plants and also, it is less labor intensive. Micro-irrigation can be beneficial in avoiding run-off on the areas which are hilly or sloped. Micro-irrigation is cost-effective. In the cost of leveling the land for any other type of farming, it is possible to install a proper micro-irrigation system.

Along with being cost effective, the micro-irrigation system can help control erosion. Micro-irrigation can prove to be helpful with direct watering methods on the farms where the crops are grown spaced closely together, such as strawberries. Also, micro-irrigation can help in controlling the water flow for the crops that are grown in the covered areas.

1.3. THE HISTORY OF MICRO-IRRIGATION

In the modern times, micro-irrigation is quickly climbing up the trend ladder and has turned out to be the standard irrigation method for many applications. Especially when it comes to home gardens and landscapes, greenhouses, vineyards, row crops and orchards, different types of micro-irrigation have been adapted.

There have been some significant changes in the technology that was being used in the initial stages of micro-irrigation. With the passage of time, the basic concept of micro-irrigation has remained constant, but there has been a significant development in the materials that are being used for micro-irrigation in the modern world.

Since the ancient times, several innovative methods of micro-irrigation have been practiced. These innovative methods of micro-irrigation were used for various other purposes than getting water onto the ground. One of the methods of micro-irrigation that was used in earlier times was burying clay pots filled with water within a planting area. Clay pots, that were filled with water, allowed the water gradually seep down into the soil and reach the zone where the roots of the plant are present.

According to some researchers, the era of micro-irrigation started in Afghanistan around the year 1866. Several reports have shown that the researchers in Afghanistan started using clay pipe to irrigate the field along with draining the planting areas.

In the era of 1920, the plantation growers and cultivators in the region of Germany started and promoted the use of perforated pipe. These perforated pipes were used to irrigate plants and drain the plantation areas.

After the development of plastics, it was used extensively after the WWII era. During this period, an inventor named Hannis Thill, of Australian origin, started promoting the use of a specific plastic pipe arrangement. This arrangement had long plastic pipes arranged in the form of passageways which lead to the even distribution of water to all the plants in the plantation area.

In the year 1959, a team of scientist named Simcha Blass and Kibbutz Hatzerim developed and patented one of the first models of practical surface drip irrigation emitter. The concept of this practical surface drip irrigation emitter was initially developed by the scientist Simcha Blass and his son Yeshayahu in Israel. The development of practical surface drip irrigation by Simcha and his son took place years before it was patented.

According to Henry Fayol, "To organize a business is to do every task require in the business functions and activities in order to produce effective and efficient results i.e., tools, raw material, capital and personnel's".

The main feature of this practical surface drip irrigation emitter was the presence of larger and longer passageways. The longer and larger passageways are used to efficiently irrigate the plantation area. As the water passes through extended passageways which slows down the velocity of water as it exits through the pipe.

The pipes, that are used in the conventional method, had several tiny holes. These holes are used for letting the water out for irrigation. The main issue with these holes was that they were easily blocked by the tiny particles such as dirt.

A number of devices such as micro-sprayers, bubblers, and misters, are being actively used in the micro-irrigation processes. These devices are capable of emitting the water in the plantation area in a manner which is different from the conventional methods and the methods used in drip irrigation as well.

The emitters used in the micro-irrigation allow the water to be transmitted to a wider area. These types of emitters are generally used on plants and the ground cover which has wider root zones. One of the most important type of micro-irrigation is the subsurface drip irrigation (SDI).

SDI generally uses permanently or temporarily buried dripline or drip tape. This drip tape is generally located at or below the plant roots. The tubing space, used in the SDI, has a number of emitters embedded. The emitters in the tubing are evenly spaced.

In the modern times, SDI is turning out to be popular for row crop irrigation. Especially when it comes to the areas where the water supply is limited or the recycled water is used for the irrigation purposes. It is important to carefully study all the relevant factors such as land topography, type of soil, characteristics of water supply, crop conditions, and climatic conditions.

This is so because studying these relevant elements helps in determining which kind of micro-irrigation system will be best suited for the crops. Studying these elements also helps in identifying the specific components that are needed to be installed in a specific area for a particular crop type.

Micro-irrigation system, used in the modern times, has turned out to be one of the most valued innovation in the field of agriculture all over the world. The series of innovations in the field of agriculture started with the invention of the impact sprinkler in the year of 1930s. The invention of impact sprinkler offered the first practical alternative to surface irrigation.

It has been observed that the impact sprinklers have greatly affected the irrigation in the recent past. Without the use of impact sprinklers, it would not have been possible to grow crops in the desert. Also, micro-irrigation is being extensively used as the primary method of crop irrigation in the regions which are susceptible to drought. Micro-irrigation has enabled to enhance the crop yields in virtually all environments while enabling the plantation owners and farmers to use less water.In 2012, a research scientist named as Dr. Daniel Hillel was honored with the World Food Prize. World Food Prize is awarded to all the individuals who have a major contribution in advancing and improving the quality, quantity or availability of food all over the world.

Dr. Hillel has extensively has worked in countries such as Turkey, Pakistan, Sudan, and other Middle Eastern countries. These countries face certain level of water shortage and by using the micro-irrigation technology proved to be very beneficial to all these countries. Dr. Hillel has been named as the "Father of Sustainable Water Management" in the modern world.

Along with Dr. Hillel, a number of researchers are working on making the agricultural field independent of the other industries. With the use of modern micro-irrigation technologies, the agricultural development is slowly becoming independent of industry for fuel. Proper utilization of micro-irrigation technology has helped in cutting down the levels of chemicals and pesticides to protect the crops and enhance the fertility of soil.

The micro-irrigation technology is still developing and evolving. Several developments and advancements are being made for the materials that are marked as the essential components of micro-irrigation technology. Plantation owners and farmers are becoming increasingly aware of the value of water as a resource.With the availability of resources and awareness, every individual in the world is recognizing the value of water in the future. Thus, with the help of micro-irrigation helps in the efficient usage and management of water for irrigation. Along with assisting in the managing the water resources, micro-irrigation technology has helped in creating a positive impact on the society.

1.4. MICRO-IRRIGATION SYSTEMS

Micro-irrigation is defined as the slow application of constant drips, tiny streams or miniature sprays of water above the surface of the soil or below the surface of the soil.In saving water as well as increasing WUE, micro-irrigation system is effective as compared to the other methods of irrigation. In addition to it, it also helps in reducing the consumption of water, growth of unwanted plants (that are known as weeds), cost of cultivation and soil erosion.

It has to be noted that micro-irrigation system can be adopted in all types of land, particularly the land where it is impossible to effectively make use of flooding method for the purpose of irrigation. A field is flooded with water in flooding method of irrigation.This results in various factors. These include anaerobic conditions around the root zone and in the soil, significant run-off, and deep irrigation below the root zone. Because of all this, sufficient amount of water is not supplied to the plants. Therefore, it is considered as one of the most ineffective surface irrigation methods.

Micro-irrigation system can be beneficial in many areas such as rolling topography, hilly areas, undulating terrain, barren land and regions that are having shallow soils. According to the depth, the types of soil can be categorized as shallow soil (that is depth less than 22.5cm), medium deep (that is 22.5–45cm) and deep soil (that is more than 45cm).

1.4.1. Features of Micro-Irrigation System

With the help of pressurized piping system, water is applied. Micro-irrigation needs pumps for developing the needed pressure for the delivery of water through pipelines, irrespective of the source of water whether it is surface or it is underground.

In case of drip irrigation systems, water is applied drop-by-drop for a long period of time;

- In order to maintain the optimum air-water balance inside the root zone, water is applied at a low rate;
- Also, water is applied at frequent intervals according to the needs of the plants;
- Water is directly provided to the plants and it is not just supplied to the other areas of the field. Therefore, decreasing the wastage; and
- Water content trapped in the soil is always maintained at 'field capacity' of the soil. Therefore, the growth of the crops is at a faster rate. Moreover, this growth rate is also consistent and uniform.

Field capacity is described as the moisture or the content of water that is present in the soil after the extra amount of water has drained away and there is decrease in the rate of downward movement. This generally takes place within 2 to3 days after irrigation or a spell of rain.

This means that the large soil pores are filled with both as well as and water after the drainage stops. On the other hand, the smaller soil pores are still filled with water. At this particular point or phase, the soil is said to be at field capacity. Moreover, it is considered to be ideal for the growth of the crop.

1.5. CLASSIFICATION OF MICRO-IRRIGATION SYSTEM

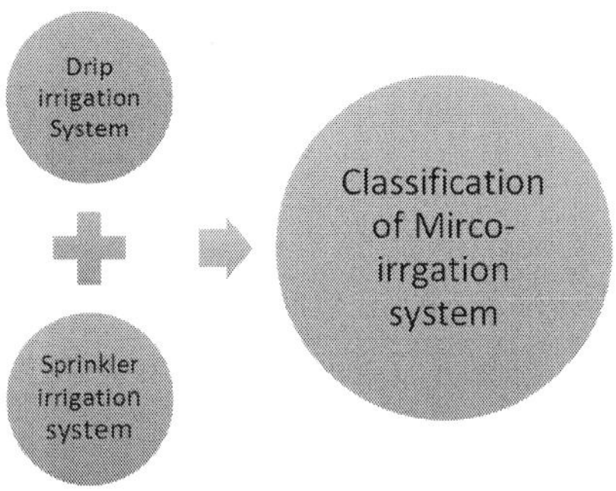

Figure 1.2. Drip irrigation system and sprinkler system: Classification of micro-irrigation system.

Micro-irrigation system can be generally categorized into two classes:

1. Drip irrigation system; and
2. Sprinkler irrigation system.

Although, it has been observed that there are a number of variances in the water flow rate, pressure requirement at which the entire irrigation system is operated and measurement of the wetted area between drip and sprinkler irrigation systems(Figure 1.2).

Water flow rate is defined as the amount of water that is discharged in a particular area at a specific time. Water flow rate expressed in gallons/ minute (gpm) or it is also expressed in liter/minute (lpm). Compensation for pressure losses must be done by the system operation pressure. It is done through system components as well as field elevation effects.

1.5.1. Drip Irrigation System

Figure 1.3. Drip irrigation is majorly used for watering the roots of the plants.

Source: Image by Flickr.

Drip irrigation system is also called as trickle irrigation system. It is defined as a technique or method of applying the amount of water, that is required, directly to the root zones of plants by the help of drippers or emitters at frequent intervals (Figure 1.3).

In drip irrigation system, water is applied by a micro jet or drop-by-drop on the surface of the soil or sub-surface. The water is applied at a rate lower than the infiltration rate of the soil. The emitters help in dissipating the pressure from the distribution system.

This is done by the means of vortexes, orifices, and tortuous or long flow paths. Therefore, it allows a restricted quantity of water to be discharged. It has been observed that most of the emitters are placed on ground, but these emitters can also be buried.

The water, that is emitted, moves within the soil system mainly by unsaturated flow. The water present on the surface of the soil wets the root zones of plants vertically by gravity. Also, it wets the roots zones of the plants laterally by capillary action.

It has to be noted that the lateral movement of water under the surface is greater in medium to heavy soil as compared to that of the sandy soil. The soil area, that is wetted, for broadly spaced emitters will, usually, be elliptical in shape. In addition to it, it has to be noted that drip irrigation can be used on windy days and it can also be used during several land operations.

1. **Types of Drip Irrigation System**

There are various classifications of drip irrigation system:

i Sub-surface drip irrigation;

ii Online drip;

`iii Family drip;

iv Surface drip irrigation; and

v In-line drip.

2. **Surface Drip Irrigation**

Surface drip irrigation is used in order to irrigate perennial crops (these are the types of plants that live for more than two years) and annual crops (these are the plants that germinate, produce flowers, seeds, and these plants die in a single year). Surface drip irrigation system include the following units:

Pump Unit: It consists of a pump and a power unit. This is used to provide electricity to the pump. The pump is used in drawing the water from the source and it helps in providing the right pressure for its delivery into the pipe system.

Head Control Unit: It includes shut-off, air, and check (non-return) valves. These are used to manage the discharge as well as the pressure of water in the whole system. There is the installation of pressure relief valve after the pump unit to return excess amount of water when the system is not functioned at its full capacity.

In addition to it, there may be filters which can be used to clear the water. The filters help in removing sediment and debris. Sediment and debris can clog the entire system. Commonly, the disc filters are used to filter water from reservoirs, tanks, ponds, and other sources of water that have algae. Also, it has been observed that some of the head control units consist of a fertilizer or nutrient tank. These are used to provide fertilizer solutions to plants.

Tubing: In tubing, a main line, sub-main lines or sub-mains and laterals is included. The main line in tubing helps in conveying the water from the source and the water is distributed to the sub-mains. The sub-mains help in conveying water to the laterals. In turn, the water is supplied to the emitters or drippers. Usually, the laterals are 13–32mm in diameter and these are used in supplying water into fields via the head control unit.

Emitters or Drippers: In order to control the discharge of water from the laterals to plants, emitters or drippers are used. Emitters or drippers are made of High-Density Polyethylene (HDPE) plastic. In drippers, the water is entered at about 1kg/cm2 pressure and the water is delivered at zero pressure in the type of droplets at a low rate of 1–2.4liter/hour. Mainly, there are two categories of emitters.

- **Online Emitters**: These are defined as the small plastic devices which help in conveying the small streams of water from polyethylene (PE) tubing to the soil. Then, the water moves through the soil through the capillary flow and it makes a wetted circle. The size of this is dependent on the type of the soil, flow rate and it is also dependent on the irrigation schedule.

It has to be noted that the online emitters are attached to the polyethylene (PE) tubing wall. These are attached by inserting the barb-shaped base of the emitter through a punched hole. It has to be kept in mind that these can be positioned at any place along the length of the pipe. In addition to it, there are some emitters that have self-piercing barbs. The diameter of pipes that is used for installing online emitters is generally between 12 mm and 20 mm.

- **In-Line Emitters or Drip Lines**: These includes small plastic emission devices. These devices are pre-inserted into the polyethylene (PE) tubing at specified intervals at the time of the tubing extrusion process. Their rate of water flow is dependent on the inlet pressure. The water flow decreases with the lower inlet pressure. On the other hand, the water flow increases with the high inlet pressure. It has been observed that this kind of emitter is available in 0.8 lph to 4 lph discharge rate.

Generally, the surface drip irrigation system is used in order to irrigate high-value vegetable crops. These crops include broccoli, celery, spinach, tomato, cauliflower, kohlrabi, leaf lettuce, etc.

3. **Sub-surface drip irrigation**

Figure 1.4. Buried plastic tubes are used in sub surface drip irrigation system for irrigating the crops.

Source: Image by Wikimedia Commons.

Sub-surface drip irrigation is defined as a technique or method of irrigating crops through buried plastic tubes. This includes embedded emitters that are placed at regular spacing's. It has to be noted that aSDI system has a same design as that of surface drip irrigation system. But in this scenario, the drip tubes are generally placed 38–84″ (97–213cm) apart and 6–10″ (15–25cm) beneath the surface of the soil (Figure 1.4).

Evaporation is minimized in case of sub-surface drip irrigation and water is used in a more effective manner as compared to that of surface irrigation. The impacts of surface infiltration such as crusting, loss of water by the means of evaporation and surface run-off are eliminated in case of sub-surface irrigation. In this system, the water is directly applied to the root zone of a crop as compared to the surface irrigation, in which most of the weed seeds hibernate.

In this system, water application is not efficient and uniform. Sub-surface drip irrigation is beneficial in many ways. It helps in water conservation in open field agriculture. Often, this results in saving up to 25% to 50% water as compared to that of the flood irrigation system.

4. **Family Drip or Gravity Fed Drip Irrigation:** Family drip or 'gravity fed drip irrigation' system is defined as a low-cost system that is developed for small family plots. This system of irrigation is best suitable for peri-urban agriculture and house gardening. In addition to it, this system of irrigation can also be used to demonstrate the working of drip irrigation system.

Family drip system or gravity fed drip irrigation system is designed for areas measuring 500–1000 m^2. There are five components in this system of irrigation. It includes shut-off valve, elevated tank, main line filter and drip line. A family drip irrigation system generally consists of a drum, main line, drip laterals and control or shut-off valve, filter (small disc or screen filter).

It has to be noted that the drip outlets are spaced at 30 cm. There is no requirement of central pressurized water system or power source in this system. Thus, this system is cheap. In addition to it, it is also easy to install and operate family drip system or gravity fed drip irrigation system.

5. **Online Drip Irrigation:** In online drip irrigation system, drippers or emitters are fixed externally on the laterals. These are placed at designed spacing's. Therefore, the drippers or emitters can be checked and in case of clogging, these can be easily cleaned.

It has to be noted that the dripper spacing can be changed at any point of the time. This is changed in order to cover the increased root zone of a plant. In vineyards, orchards, artificial landscapes and nurseries, online dripper system is used.

Generally, this system is used for irrigating horticultural plants such as orange, lemon, banana, papaya, guava, bamboo, mango, coconut, grapes, pomegranate, sapota, teakwood, amla (Indian gooseberry), etc.

6. **In-Line Drip Irrigation:** Drippers are fixed in the lateral tube in the in-line drip irrigation system. Drippers are fixed at designed spacing's at the time of manufacturing in order to meet the needs of several crops. This system is effective for row crops such as sugarcane, groundnut, cotton, vegetables, and flowering crops. Dripper spacing is dependent on the requirement of water of a crop. In addition to it, it is also dependent on the water holding capacity of the soil. Once this system is installed, no change can be done in dripper spacing.

1.5.2. Sprinkler Irrigation System

Figure 1.5. Sprinkler irrigation system sprays the water all over the plantation area just like rain.

Source: Image by Flickr.

Sprinkler irrigation is defined as a technique or method that is used for applying water in a way that is similar to rain. This system is best suited for most row, field crops as well as tree crops. In this system, water can be sprayed over the crop canopy or under the crop canopy. The rate at which the water is sprayed over the crop or under the crop canopy is measured in mm/hour (Figure 1.5).Sprinkler system will not be right or suitable for the place that is known to be windy most of the time. The water is broken down into droplets sized 0.55mm to 4mm by the sprinkler. The size of the drop is controlled by pressure as well as by the nozzle size of the sprinklers.

The application rate is dependent on the operating pressure, the size of sprinkler nozzles and distance that is between the sprinklers. It has to be noted that the application rate must not be more than the maximum permissible infiltration rate for a particular soil type.

If the application rate exceeds, it will result in loss of water, soil erosion and surface sealing. In addition to it, there may be insufficient moisture in the root zone of crops or in the plants after irrigation. Moreover, it can also provide damage to the plants.

Water pressure is defined as the force with which the water flows out of the sprinkler. It is measured in pounds per square inch (psi). Therefore, sprinklers are designed in order to work at certain levels of pressure, which are suggested as their operating pressure.

The distribution of water will be affected if the pressure is below or above than the recommended level of pressure. The water drops become larger when the pressure is low. And they cannot irrigate those crops that are far away from the system.

On the other hand, if the pressure is high, then the water drops will be smaller, and they will not irrigate the crops in an even manner. Because of this the sprinkler heads can also be damaged. Sprinklers are best suited for sandy soil even though they are adaptable to most soils. In addition to it, sprinklers can be used for gardens, irrigating lawns as well as agricultural fields.

Sprinkler irrigation system have the following types:

i. Centre pivot;
ii. Rain gun;
iii. Impact sprinkler;
iv. Pop up sprinkler;
v. Towable pivot; and
vi. Linear move sprinkler.

1.6. OTHER TYPES OF MICRO-IRRIGATION SYSTEM

1.6.1. Bubbler Irrigation

In order to irrigate larger areas and apply water on a 'per plant' basis, bubblers are used. Water from the head of the bubbler either runs down from the emission device or the water spreads a few inches in the pattern of an umbrella. In bubbler, the emitters dissipate the pressure of water through a variety of diaphragm material (it is a silicon diaphragm inside an emitter flexes to control water output). The water is deflected through small orifices. It has to be noted that the bubbler emission devices are fitted with single pot outlets or multiple port outlets. In addition to it, bubblers are also available in adjustable flow as well as in pressure compensating types.

1.6.2. Spray Irrigation

In spray irrigation system, foggers, jets, or misters are used. These are also known as 'spitters.' In this system, water is applied just to a portion of the ground surface. Although, in spite of dripping water from narrow orifice emitters, micro-sprayer systems eject fine jets of water. These 'jets' of water usually come out from the series of nozzles. Every single nozzle is capable of watering an area of several square meters. This tends to be much larger area than individual areas that are wetted by a single drip emitter.

1.7. MERITS OF MICRO-IRRIGATION OVER OTHER SYSTEMS OF IRRIGATION

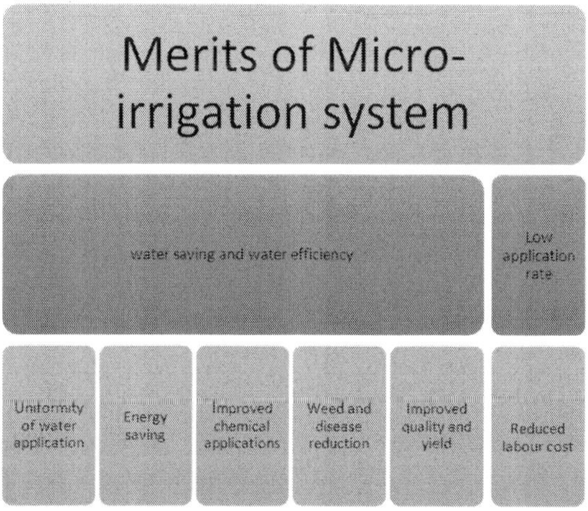

Figure 1.6. Some of the advantages of micro-irrigation system.

There are various advantages of micro-irrigation systems when it is compared to other methods of irrigation (Figure 1.6):

1. **Water Savings:** The requirement of irrigation water can be much smaller when it is compared with other methods of irrigation. This is because of the irrigation of a smaller part of the soil volume, reduced evaporation from the surface of the soil and the decrease or removal of the runoff. Since the micro-irrigation system permits for a high level of water control application, water can be used only when it is required and deep percolation can also be reduced or it can be prevented.
2. **Water Efficiency:** Micro-irrigation can help in reducing the usage of water by 25% to 40% as compared to the overhead systems, and can reduce 45–60% as compared to surface irrigation.
3. **Low application rates**: With the low application rate, it is meant a less expensive irrigation system and more effective use of pumps, pipelines, and filters. This is because the components of

these system may be sized for lower flow rates and can be utilized for longer periods of time. Micro-irrigation systems are designed in order to supply the water requirement of an individual plant by a day-to-day application.

4. **Uniformity of Water Application:** In micro-irrigation systems, there is excellent uniformity of water application; thus, all the plants get the equal amount of water. The results of good uniformity are more efficient irrigation. Further, this leads to savings of power, water, and fertilizer. An even andconsistent use of water also results in more uniform and better yields. This is because every single plant is provided exactly as much water as well as nutrients as it requires for optimum growth.
5. **Energy Saving:** There is the need of a smaller power unit compared to other systems of irrigation. Generally, the delivery pipe systems function under low pressure (that is 2 bar - 4 bar) and hence there is the need of less amount of energy for pumping.
6. **Improved Chemical Application:** Micro-irrigation systems permits for a high level of control of chemical applications. Every single plant can be provided with the exact quantity of fertilizer that is needed at a particular time. While they are applied directly to the root zone, a decrease in the total quantity of fertilizer that is used is possible (average 25% to 50% cost savings in chemicals as well as fertilizers). It has to be noted that this application method is more efficient.

It helps in providing better distribution of nutrients all through the season and reduces the ground water pollution because of the high concentration of chemicals that could normally move about with deep percolated water. There are various other chemicals like insecticides, fungicides, herbicides, growth regulators and carbon dioxide that can be efficiently applied through micro-irrigation systems in order to enhance the production of the crop.

7. **Weed and Disease Reduction:** The growth of weed is inhibited, and disease occurrences reduced because of the limited wetted area.
8. **More Flexibility in the Field Operations:** Micro-irrigation technology can be applied on windy days. During the watering operations, the micro-irrigation system can function without interruption when harvesting.

9. **Improved Tolerance to Salinity:** Micro-irrigation helps in decreasing the sensitivity of most of the crops to saline water, soil-water conditions because of the maintenance of high levels of moisture in the root zone. The common application of water constantly substitutes moisture removed by the plant and it also moves salts to the edges of the root zone that is away from the plant.

These salts precipitate out and accumulate at the edge of the wetted parameter. This process helps in preventing the damaging combination of high soil salinity and low moisture from happening. Thus, those crops that are under micro-irrigation systems are more tolerant of saline water as well as the soil conditions.

10. **Improved Quality and Yield:** The quality of crop and yield under micro-irrigation is enhanced. This is because of the regular, slow, uniform application of water as well as nutrients. In addition to it, all the harm and losses because of the water contact with fruit or foliage are also eliminated.

11. **Adoption to Any Topography and Soils:** Micro-irrigation systems can function in an effective manner on any topography if it is designed and managed properly. With micro-irrigation system, the low application rate is considered perfect or ideal for heavy clay soils that are having low infiltration rates as the enough quantity of water can be applied slowly for the soil to absorb it without croding the surface.

It has to be noted that very sandy soils, on the other hand, frequently cannot store large quantities of water. For such kind of oils, micro-irrigation is ideal too. This is due to its ability to regularly provide small quantities of water to the crop.

12. **Automation:** The micro-irrigation system can be automated easily by the use of electrical solenoid valves as well as controllers. This permits the system to be functioned at any time of the day or during the night and for any desired length of time. Thus, this allows the irrigation managers to take the benefits of information related to crop water use in determining the optimum irrigation time.

13. **Reduced Labor Cost:** Labor savings are one of the main advantages of the micro-irrigation system. It has been observed that the labor requirements are low because the micro-irrigation

system allow larger areas to be irrigated at the same time and because the systems can be fully automated. Apart from the direct savings in labor, often there are indirect labor savings because of the decreased number of cultivations, the removal of the application of fertilizer as a separate operation.

It has been observed that most of the sprayers, sprinklers, and jets are insect protected nozzles. After the operation, they are closed in order to avoid any clogging that is caused by insects or any other debris. If all these advantages of micro-irrigation benefits are not achieved, the investment that is done in a micro-irrigation system is not of any value or worth. It has been observed that product life that are having equipment of good quality, good operation as well as management can last up to 15 years to 20 years(Figure 1.7).

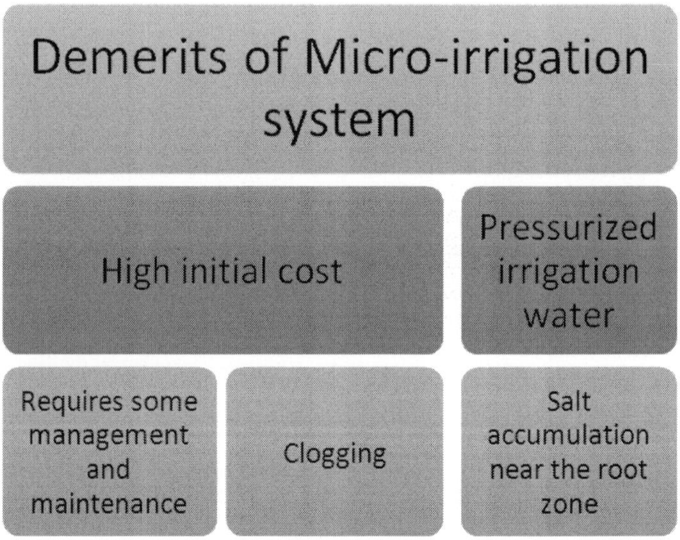

Figure 1.7. Some of the shortcomings of micro-irrigation system.

1.8. DEMERITS OF MICRO-IRRIGATION SYSTEMS

1. **High Initial Cost:** It has to be noted that the initial investment as well as the cost of maintenance for a micro-irrigation system maybe higher as compared to some other methods of irrigation, but it is the responsibility of the growers to weight the cost against the advantages.

There are various components such as pumps, regulators, valves, Filters, gauges, chemical injectors and possible automation components that add to the entire cost of a micro-irrigation system. The emitter itself (sprayer, drip tube/tape, and sprinkler) indicates just approximately 35% to 37% of the initial system cost. Actual cost will differ considerably. This depends on the selection of a specific micro system.

The growers must also understand that a well-designed, installed as well as managed micro-irrigation system has water saving and many other significant environmental, agronomic, and economic advantages.

2. **Pressurized Irrigation Water**: It is necessary to pressurize the irrigation water. This results in energy costs. The pressures that are required are usually less than those that are required for sprinkler systems, but these pressures are higher as compared to those of the flood irrigation systems.
3. **Requires Some Management and Maintenance:** Generally, farming with micro-irrigation systems needs a change in various aspects such as cultivation, planting as well as harvesting practices. It is very important to educate growers about these changes before and also during the first season of the micro-irrigation. All these latest or new practices can rapidly become a standard part of the farming operation.

Generally, the micro-irrigation systems have greater maintenance. This is because of the small orifice features that are susceptible to clogging. This clogging is caused from organic matter, particulate matters, and chemical precipitates. Thus, there is the need to have additional maintenance which includes injecting chlorine or acid, filtration, and flushing lateral lines. This is required in order to ensure best performance.

There are many factors that can cause leaks in the drip tape. These includes food traffic, machinery, animals, insects in the field. In addition to it, in order to understand the various advantages that are mentioned above, it is the duty of producers to constantly monitor the increasing environment and scheduling irrigation in order to meet the needs of the plant.

4. **Clogging:** Clogging of emitters is one of the biggest problems that is encountered in micro-irrigation system. It has been observed that the small openings can be clogged easily by the particles that are present in the soil, bacterial slime, organic matter, algae or chemical precipitates. There is the need of a very good filtration (200 mesh filtration degree is most often

recommended) along with the good quality water supply in micro-irrigation system. The filtration system should be selected based on chemical characteristics, physical characteristics or biological characteristics of the water.

5. **Salt Accumulation Near the Root Zone:** It has been observed in case of surface and sprinkler irrigation systems the salt is moved below the crop root zone. In case of micro-irrigation systems, salt is moved to the outer edge of the wetted volume of soil and surface of soil surface. In addition to it, inadequate rainfall can flush the salts back into the root zone and it can provide harm or damage.

There is a need for careful management in order to make sure that the salts do not go back into the active root zone. A sprinkler or surface irrigation system may have to be used if the need to leach salts from the root zone becomes critical. This purpose is accomplished in an effective manner with the help of sprinkler or surface irrigation. In all those places or areas which are having heavy rainfall, the salts will be washed out of the root zone prior to the occurrence of significant accumulation.

6. **Seed Germination:** It has been observed that some of the crops do not germinate well with micro-irrigation systems (normally under drip tube or drip tape). In such kind of situations, portable sprinklers are often used for the process of germination. Once the seed germinates and starts growing, the crop can be irrigated with micro-irrigation to optimize plant growth.
7. **Moisture Distribution/Restricted Root Zone:** Moisture distribution is largely dependent on the soil type that is being irrigated by the use of micro-irrigation system. It has been observed that in some soils, that is deep sands, little lateral water movement (that is low capillary forces) can give rise to many of the issues.

Under such situations, it is very hard to wet a significant part of the root zone. In addition to it, without deep percolation, it is even more difficult to manage the irrigation since only a small quantity of water can be kept in the wetted volume desired.

The increase in the number of emitters per plant may help in improving the distribution of water in the soil. Thus, coarse sands will need much closer spacing of emitters as compared to the fine soils. Generally, for any kind of soil, the number of emitters as well as their spacing must be grounded on the geometry of wetted soil volume.

It has been observed that plant root activity is often restricted to the soil zone wetted by the emitters mainly in the areas with low rainfall. It is very important for the irrigator to remember that the micro-irrigation system is meant to apply frequent and small irrigations. It has to be noted that the cover crops cannot be grown all over the year because of the localized nature of the water applications.

1.9. CONCLUSION

Since the dawn of life on planet Earth, human beings have been trying to improve their food cultivation and agricultural skills. With constantly changing environment and unprecedented development of technology, human beings regularly face different problems and have been discovering or inventing methods and equipment that can enhance the yield of the crops. In the modern times, one of the major issues is the unavailability or depletion of natural resources.It has been observed that most of the countries in the world are on the verge of facing scarcity of one of the vital elements that helps in sustaining life-Water.

Farmers and plantation growers in the world are focusing majorly on management of the natural resources, especially water, to maximize the crop yield. So, in order to effectively manage water along with other resources, many scientists and experts developed micro-irrigation.

Micro-irrigation technology has proved to be very helpful in the recent times when it comes to managing the natural resources. Along with effectively spreading the water through the plantation areas, micro-irrigation has been very effective in holding the soil particles together and preventing the soil erosion.Micro-irrigation has two major divisions. Drip irrigation focuses on watering the root zones and providing all the essential nutrients to the plant roots. On the other hand, Sprinkler irrigation sprays the water in the form of rainfall. This is very effective in the areas where the soil particles are unable to hold the water. One of the most remarkable features of micro-irrigation technology is that it is sustainable. It does not have any water rundown system. So, it prevents soil erosion while effectively watering the plant.

With technological advancements, the micro-irrigation technology has been constantly evolving. There is a great need to assist and promote the micro-irrigation technology all over the world so that water can be used effectively. Most importantly, micro-irrigation technology is sustainable and would definitely help in enhancing the crop yield while promoting the effective usage of the resources.

REFERENCES

1. (2019). Introduction to Micro-Irrigation System. [ebook] pp. 1–42. Available at: http://ncert.nic.in/vocational/pdf/kvmt101.pdf (accessed on 20 August 2020).
2. Drip Depot DIY Irrigation Support, (2020). The History of Drip Irrigation. [online] Available at: https://help.dripdepot.com/support/solutions/articles/11000044434-the-history-of-drip-irrigation (accessed on 20 August 2020).
3. Ecoursesonline.iasri.res.in. (2020). MI&CAD: Lesson 1. Introduction to Micro-Irrigation. [online] Available at: http://ecoursesonline.iasri.res.in/mod/page/view.php?id=196 (accessed on 20 August 2020).

Chapter 2

Solar Energy: A Fundamental Review

CONTENTS

The chapter of explains the basic significance of the production of electric energy with the help of the solar radiations. This chapter also explains how the conversion of solar radiation into the form of electricity or electric current takes place.

This chapter provide highlights on the several numbers of various uses of solar energy such as thermal energy, electricity generation, solar water heating, solar ventilation, solar lighting, portable solar, solar transportation, wearable solar tech, along with the other applications.

This chapter also explains the several components of the solar power systems such as the roof system, solar panels, inverter, and net meter.

This chapter also highlights on factors which need to be considered before the installation of the solar power systems, along with this, the chapter also mentions the impact of solar energy on the environment.

This chapter also addresses the importance of the solar energy such as the promise of the solar energy, the growth of the solar energy, the basics of the solar energy. This chapter also includes the advantages and disadvantages of the solar energy.

2.1. INTRODUCTION

The definition of the solar energy is, the radiation that is coming from the sun, and having the potential of producing heat, triggering the chemical reactions, or producing the electric energy.

The total amount of the solar energy that strikes the Earth is massively in excess of the present as well as expected necessity of the energy, all across the globe.

It is suitably harnessed, the highly diffused source has the potential in order to fulfill the requirement of the energy in the times to come

Talking about the 21st century, the solar energy is anticipated in order to become more and more attractive in the form of the source of renewable energy.

And the reason behind this is its never-ending supply and its non-polluting property, in stark contrast to the finite fossil fuels coal, natural gas, and petroleum (Figure 2.1).

Figure 2.1. An image of solar panel.

Source: Image by Pxfuel.

The sun is a highly powerful source of the energy, and the solar radiations is by far the largest source of the energy that is received by the Earth, but in addition to this, the intensity of the solar radiation or the sunlight at the surface of the Earth is actually quite low. This is mainly due to the enormous radial spreading of the radiation from the huge distance of sun.

A comparatively minor additional loss is because of the atmosphere and the clouds of the Earth. The atmosphere and the clouds absorb or scatter as much as 54% of the incoming light from the sun. The solar radiations or light from sun reaches the ground is consisting of near about 50% of visible light, near about 45% of infrared radiations (IR), and little amount of the Ultra-violet (UV) rays, along with the other kids of electromagnetic radiations.

The sun does more than for the planet Earth in spite of just providing light through the course of the daytime. In which, each, and every particle of the solar radiation or the sunlight (that is known as photon) that reaches the Earth, and is consisting of energy that fuels the Earth.

Solar energy is the ultimate source, that is accountable for all of the weather systems and energy source that are found on the Earth, and sufficient amount of sunlight reaches the surface of the Earth each hour in order to theoretically satisfy the global energy requirement for near about a whole year.

Where does all of this energy come from? The Sun, which is just like any kind of star that is present in the galaxy. The sun is like a huge nuclear reactor. Deep in the core of the Sun, due to the nuclear fusions reactions huge amount of energy generates that radiates in the upward direction from the surface of the Earth as well as into the space in the form of light and heat.

Solar power can be harnessed and can be transformed in the form of usable energy with the help of the PV (photovoltaics) or solar thermal collectors. Nevertheless, solar energy is only responsible for a little amount of overall consumption of energy, the falling cost of installing solar panels means that more and more numbers of people in several places can have the benefits of solar energy. Solar is a resource of renewable energy, it is clean, and in the end, it plays a significant role in the global energy future.

2.2. WHAT IS SOLAR ENERGY?

As per the scientific observations and researches, the Earth receives huge amount of solar energy or solar power, to be specific it is one hundred seventy-three trillion terawatts. That is literally ten thousand time more power when it is compared with the amount of energy consumption by people. This proves the fact that the Sun is the most abundant source of the energy on the entire globe and that it could one day be the most needful source of the energy.

Conventionally, all across the globe, the electrical requirements have been satisfied with the help of the fossil fuels such as coal, oil, natural gas, and so on. Nevertheless, these sorts of energy sources have two negative effects:

This plays a significant part in the global warming as well as the acid rain pollution. As an outcome, it negatively affects the several numbers of various animals, plants, along with the humans in the environment.

Several nations have full access to the energy resources that are based on the fossil fuels, which can take into the direction towards the global political as well as economic stability (Figure 2.2).

Figure 2.2. An illustration of solar panels installed on the rooftop.

Source: Image by Pxfuel.

The best substitute is the solar energy, which is a renewable resource. In simple words, solar energy will not become unavailable at any point of time. Solar energy offers a never-ending, steady supply across time. Solar energy is also green source of energy, and the reason behind this that, solar energy does not emit pollutants through the course of the process of energy production.

Solar energy is the energy that is generated with the help of the Sun in the form of heat and light. Solar energy is one of the most renewable and readily available sources of the energy that exists on the Earth. As a matter of fact, solar energy is available in huge amount and it is free, as well as solar energy does not belong to any person, which makes the solar energy one of the most significant of the non-conventional sources of the energy.

Solar energy has been used with the help of people, from the ancient times simply with the help of magnifying glass in order to concentrate the sunlight in the form of beams, in such a way that they would cause wood in order to catch fire.

Necessarily, solar energy can be utilized in order to transform it in the form of heat energy or the solar energy can be transformed in the form of electric current. Solar energy is the energy that is harnessed from the sun.

2.3. CONVERSION OF SOLAR ENERGY

There are several numbers of various methods in order to utilize the energy or the solar energy that is coming from the Sun. Out of which, the two main methods in order to utilize the energy from the sun are Photovoltaics (PV) and second is, solar thermal capture.

PV are comparatively more common for the smaller-scale electricity projects (such as installation of the residential solar panels). On the other hand, the solar thermal capture is generally only used with respect to the production of electric current on huge scales in utility solar installations.

Moreover, to the production of the electric current, lower temperature fluctuations of the solar thermal projects can be used for heating and cooling as well.

Solar energy is the one of the rapidly growing as well as cost effective source of the power in the world, and it will continue to spread with much more rapid pace in the coming interval of time. With each passing year, the improvements in solar panel technology, the economic benefits of the solar improve, adding to the environmental perks of choosing a clean, renewable energy source.

2.3.1. How Solar Energy Is Converted to Electricity?

The first and foremost step, in order to transform the solar energy in the form of electric current, is the installation of PV cells (or photovoltaic cells) or solar cells. The term "photovoltaic" means light and electricity. These solar cells capture the energy coming from the Sun and then transform it in the form of electric current (Figure 2.3).

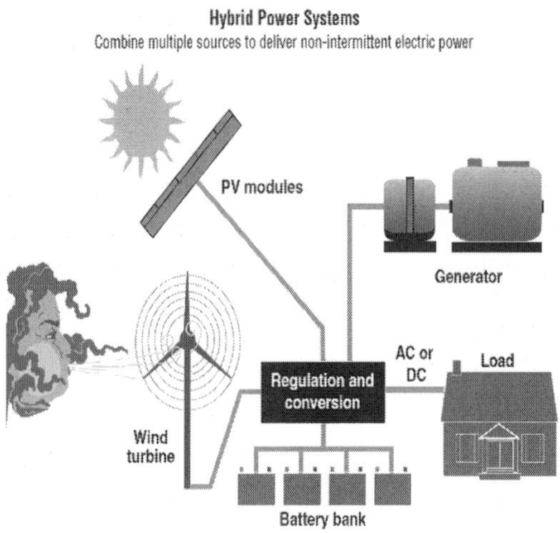

Figure 2.3. An illustration of conversion of solar energy: A hybrid power system.

Source: Image by Wikipedia.

The solar cells or PV cells are made up with the help of material that depicts the photovoltaic effect. In simple words, when the solar radiations reach the PV cells, the photons of the light activate the electrons that are present in the solar cells, triggering them in order to start flowing, and in the end, the production of electric current takes place.

When looking to purchase the solar panels, it would be very helpful in order to know the options that are available in the market. Here is an outline of the main ones, as mentioned below:

1. Polycrystalline-This utilizes Multi crystalline Silicon;
2. Mono-crystalline-This is ideal for small spaces; and
3. Thin Film-Generally, bigger in size and a lot more effective and successful through the course the day.

The difference between the material that is being used in order to manufacture the mono-crystalline and polycrystalline lies in the composition of the silicon substrate which is used to manufacture the PV cells and, in the end, the solar panels. As the name implies, polycrystalline point toward several numbers of crystals, on the other hand, mono-crystalline point toward a single crystal.

As the size of the crystals are huge, the solar cell or the PV cell will be more effective, which explains the reason mono-crystalline cells are near about 10% to 15%, when it is compared with the polycrystalline crystals.

2.4. USES OF SOLAR ENERGY

The capabilities of the solar energy are massive, since near about two hundred thousand times the total daily electric-production capacity of world, is received by the Earth with each passing day in the form of solar energy.

Never the less, the solar energy itself is free for all, the high cost of its collection, conversion, as well as the storage still restricts the usage of the solar energy in several numbers of different places, all across the world.

The solar radiation or light coming from the Sun can be transformed either in the form of thermal energy (which is heat) or in the form of electrical energy (which is electric current), nevertheless, the former is much easier to achieve.

2.4.1. Thermal Energy

Out of the most commonly devices that are being used in order to harness the solar energy and transform the solar energy in the form of thermal energy or heat, are flat collectors, which are used for solar heating applications. The reason behind this is the intensity of the solar radiation at the surface of the Earth is so low, these flat collectors must be large in area in order to fulfill the process more effectively.

Even in the regions of the World, where the solar radiation or sunlight found in plenty amount or in sunny days, for example, a collector must have the surface area of near about forty square meters (which is nearly four hundred and thirty square feet) in order to collect the sufficient amount of energy to fulfill the energy requirements of an individual.

The most extensive used flat-plate collectors are consisting of a blackened metal plate, that is covered with one sheet or two sheets of glass, that is heated with the help of solar radiation or sunlight falling on it.

This heat is then conveyed to the air or water, which are known as carrier fluids, that flow past the back of the plate. The heat can be used directly, or also, the heat can be transferred to the other medium of storage. Generally, the flat-plate collectors are used for solar water heaters and house heating.

The storage of the heat for use at night or on cloudy days is generally achieved with the help of using the insulated tanks in order to store the water heater through the course of the sunny days.

These kinds of systems can provide a home with hot water that is drawn from the storage tank, or, with the warmed water flowing across the tubes in floors and ceilings, it can provide space heating.

Generally, the flat-plate collectors are the heat carrier fluids to the temperature which is ranging from near about 66°C to 93°C (nearly 150 degrees Fahrenheit to 200 degrees Fahrenheit).

The effectiveness of these sorts of collectors (for example, the amount of the energy received that they transformed in the form of usable energy) ranges from 20% to 80%, generally relies on the design of the collector.

Apart from this, the other method of the transformation of thermal energy is found in the solar ponds, which are the bodies of salt water that is particularly designed in order to gather and store the solar energy.

The heat, which is extracted from these kinds of ponds allows the generation of chemicals, chemicals, food, textiles, along with the other industrial products and also, can be used in order to supply heat to the greenhouses, swimming pools, and livestock buildings as well.

So often, the solar ponds are used in order to generate the electric energy with the help of the use of the organic Rankine cycle engine, which is comparatively effective and cost-efficient method of conversion of solar energy, which is particularly useful in the remote areas. Solar ponds are fairly costly in order to install and maintain, and are commonly restricted to the warm rural regions.

With respect to the smaller scale, the energy coming from the Sun can also be used in order to cook in specially designed solar ovens.

Conventionally, the solar ovens concentrate the sunlight from an extensive area to a central point, where a vessel that is consisting of black surface helps in order to transform the solar radiations in the form of heat. Generally, the solar ovens are portable in size and necessitates no other kind of fuel inputs.

2.4.2. Electricity Generation

Solar radiations may be directly transformed in the form of electric current with the help of PV cells (also known as solar cells). With these kinds of cells, a small electric voltage is produced when the light reaches the junction between a metal and a semiconductor (for example, silicon) or the junction between the two various kinds of . The energy, which is produced with the help of single PV cell is commonly only about two watts.

With the help of connecting large numbers of individual cells together, nevertheless, just similar to the solar-panel arrays, hundreds or even thousands of kilowatts of electricity or electric energy can be produced in a solar electric plant or in a large household array.In the present interval of time, most of the energy effective PV cells is only near about 15% to 20%, and, since the intensity of the sunlight is low in order to start with, large, and expensive assemblies of these kinds of cells are need to generate even moderate amounts of power.

Small PV cells that works on the solar radiations or light coming from the sun, along with the help of artificial light have found major use in applications that necessitates low power, just similar to the case of power source for calculators and watches.Larger units have been used in order to supply the power for water pumps and communications systems in remote regions and for weather and communications satellites.

Classic crystalline silicon panels and developing technologies with the help of thin-film solar cells, which is consisting of building-integrated photovoltaics, can be installed with the help of the owners of the home and businesses on their rooftops in order to substitute or augment the traditional electric supply.Concentrated solar power plants use concentrating, or focusing, collectors in order to focus the solar radiations or light from the sun received from an extensive area on to a small blackened receiver, in this way, significantly elevating the intensity of the light to generate the high temperatures.The sets of carefully aligned mirrors or lenses can concentrate sufficient amount of sunlight in order to heat a target to the temperature level of near about two thousand degrees Celsius (or 3,600 degrees Fahrenheit) or more.This heat can then be utilized in order to operate the boiler, which, as an outcome, produces steam for a steam turbine electric generator power plant. In order to generate the steam directly, the movable mirrors can be aligned in such a way that they concentrate huge amount of solar radiation on the blackened pipes with the help of which water is circulated and so as heat.

2.4.3. Solar Water Heating

Application of solar energy extend to water heater systems. Most of the solar water heating solutions create hot water, which is utilized in the houses (Figure 2.4).

Figure 2.4. An illustration of solar water boiler.

Source: Image by Flickr.

Solar water heaters necessitate a rooftop cell in order to absorb the heat coming from the Sun and transfer that heat to the water tank. Generally, solar water heaters have five years to ten years payback, as per the United States Department of Energy (DOE).

Apart from this, the other application of solar energy, more particularly in the southern Unites States and south-western United States, is heating swimming pools. Water is circulated to a collector where the water is being heated with the help of the sunlight and then pumped back into the pool.

With costs between three thousand dollars to four thousand dollars and a lifetime of near about 1.5 years to 7 years. According to the United States DOE, which states that, "solar pool heating is the most cost-efficient application of solar energy in several kinds of different climates."

2.4.4. Solar Heating

General application of solar space heating systems is consisting of powering radiant floors or pairing with a Forced Hot Air (FHA) system in order to supply heat to a house. Passive solar home design can also heat the homes and businesses in the winter with the help of considering the placement of the windows and the selection of the materials that are being used in the construction.

2.4.5. Solar Ventilation

Solar ventilation solutions, for example, solar attic fans can decrease the load of the HVAC (which is short form for Heating Ventilation and Air Conditioning) with the help of cooling the home through the course of summers. Solar ventilation can be good option if an individual is not able to install a solar PV system that offsets the entire electricity use of the house. Apart from this, one innovative product is the Sola-tube solar attic fan (Figure 2.5).

Figure 2.5. An illustration of solar attic fan.

Source: Image by pxhere.

Industrial applications as well as industrial applications of the solar process heat is consisting of solar ventilation technologies that can pre-heat the air of a building in the cold climates, which decreases the energy costs.

2.4.6. Solar Lighting

In recent times, solar lights have mostly used at every place and can be found everywhere from home landscaping and security lights to the road signs and street lights. These kinds of solar lighting technologies for the homes are cost efficient in nature and readily available from basic to high-end designs everywhere from the local hardware store to e-commerce websites, for example, Amazon.com.

Apart from this, one innovative application of indoor solar lightning, feature on Mashable, in the Sola-tube skylight that adds natural light, along with the decreasing the use of energy.

2.4.7. Portable Solar

In the connected world, smartphones, and tablets are the necessities and, let's face it, every so often running low on the battery. Portable solar photovoltaic (SPV) chargers can keep the personal devices that operates on electricity, charged on the go. The technology already exists in order to combine the solar cells or PV cells into the smartphones and has been used in watches from the time period of 1970s (Figure 2.6).

Figure 2.6. An illustration of portable solar charger.

Source: Image by wallpaper flare.

Taking an example of Japan, researchers have developed light-weight, water-resistant PV cells that could someday be sewed into the clothing in order to supply the power to the devices.

2.4.8. Solar Transportation

Vehicles that operate on solar power may be the future, with existing applications that are consisting of buses, trains, airplanes, and race cars built with the help of the students in Australia and the United States. A car, which is entirely powered with the help of the solar cells or PV cells, is even slated for commercial release in the year of 2019.

This application of solar energy is not yet extensively available, unless an individual or a person owns an electric car or electric vehicle (EV) and charge it with the help of the solar panels (commonly with the help of the solar-connected battery).

2.4.9. Wearable Solar Tech

In the present interval of time, powering the consumers electronics has become a usual solar power application, for example, chargers that are solar-powered, such as Anker's Power port, can charge anything, ranging from a smart phone to a tablet or e-reader. There are even flashlights that are powered by solar energy, that can be charged with the help of just being exposed to the solar radiations or sunlight.

For those who are curious about the products in solar technology, here is the list of top ten of application of the solar devices, mentioned below:

1. Cell phones;
2. Wearables;
3. Music speakers;
4. Solar air conditioning;
5. Tablets;
6. Thermostats;
7. Solar dryers;
8. Solar visor radios;
9. Freezers/mini-fridges; and
10. Rechargeable flashlights

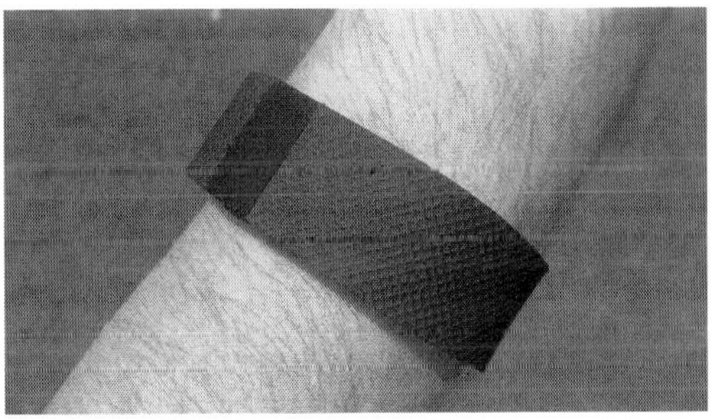

Figure 2.7. An illustration of wearable solar tech.

Source: Image by Wikipedia.

As the world move away from the use of the fossil fuels and search for renewable sources with respect to the daily basis electronics, there is no doubt that the solar energy will continue to adopt by the huge amount of people for any kind of product that can be exposed to solar radiations or the sunlight (Figure 2.7).

The main objective, utilization of the solar products in order to reinvent the methods and techniques that a person thinks of power usage, and reverse the expectation for how devices should be charged. And as the conversion efficiency of the solar endures to enhance, this trend will expand into the new markets as well as new products.

For instance, wearable solar could soon make Apple Watches, and Fitbits much more appealing to the consumers, not like to their standard counterparts, smart watches that are solar-powered do not require to be plugged into recharge every night.

2.4.10. Other Applications

Solar energy is also utilized on a small scale for purposes other than those which are mentioned in the above topics. With respect to some nations, for example, solar energy is used in order to generate salt from seawater, with the help of the process of evaporation.

In the same manner, solar-powered desalination units convert the salt water into the drinking water with the help of transforming the solar radiations in the form of heat, directly or indirectly, in order to conduct the process of desalination.

Solar technology has also developed for the clean as well as renewable production of hydrogen as a substitute source of energy. Imitating the process of the photosynthesis, artificial leaves are silicon-based devices that utilizes the solar energy in order to separate the water in the form of hydrogen and oxygen, virtually leaving no pollutants.

In addition to this, further work is required in order enhance the effectiveness as well as cost-effectiveness of these kinds of devices for industrial use as well.

2.5. COMPONENTS OF SOLAR SYSTEMS

Solar power is one of the cleanest, most reliable form of renewable energy that is available. Solar energy can be used in various kinds of forms in order to power the houses or businesses.

Solar-powered photovoltaic (PV) panels transforms the solar radiations or sunlight in the form of electric current with the help of exciting the electrons that are present in the silicon cells with the help of light that is coming from the sun. This electric current can be utilized in order to supply the renewable energy to the house.

In order to understand this process deeper, let's take a look at the components of the solar energy that compose an entire solar power system.

2.5.1. The Roof System

With respect to the most of the solar systems, the solar panels are situated on the roof. An ideal site will have no shade on the panel, more particularly through the course of the prime sunlight hours of 9 A.M. to 3 P.M. Generally, a south facing installation will give the optimum potential for the solar power system, apart from this, other orientations or other settings may provide the enough production of electricity (Figure 2.8).

Figure 2.8. An illustration of solar panels installed on the rooftop of a house.

Source: Image by Wikimedia commons.

Trees or other aspects that produce shading through the course of the day will generate considerable amount of decrement in the production of power. The significance of the shading and efficiency cannot be overstated. With respect to the solar panel, if even just one of its 36 cells are under the shading, the production of power will be decreased by more than half.

The well-known installation contractors, for example, NW Wind and solar employ a device which is known as Solar Pathfinder in order to carefully determine the potential regions of shading before the process of installation.

Not each and every roof has the correct orientation or angle of inclination in order to have the benefit of the solar radiations. That is why, some kinds of systems are designed with pivoting panels that provide assistance in order to track the sun in its journey across the sky.

Non-tracking photovoltaic systems should be inclined at an angle which is equal to the latitude of the site in order to absorb as much amount of energy throughout the year. Apart from this, alternate orientations and/or inclinations can be used in order to optimize the production of energy for particular times of the day for with respect to the particular seasons of the year.

2.5.2. Solar Panels

Solar panels which are also known as the modules. These modules are consisting of PV cells or solar cells. Solar cells are made up with the help of the silicon that help in converting the incoming solar radiations in the form of electricity, in spite of heat. ("Photovoltaic" means electricity from light—photo = light, voltaic = electricity).

Solar PV cells are consisting of a positive as well as a negative membrane of silicon which is situated under a thin slice of glass. As the photon of the light coming from the sun strikes on these solar PV cells, they knock off the electrons that are present in the silicon.

The negatively-charged free electrons are specifically attracted to one side of the silicon cell, which produces an electric voltage that can be gathered and channeled. This electric current is collected with the help of wiring the individual solar panels together in sequence in order to form a solar PV array.

According to the size of the installation, several numbers of strings of the solar PV array cables end-up in one electrical box. This electrical box is known as a fused array combiner (Figure 2.9).

Figure 2.9. An illustration of solar panels.

Source: Image by air force civil engineer center.

Contained within the combiner box are fuses which are designed in order to safeguard the each and every module cable, along with the connections that deliver power or electric current to the inverter. The electric energy produced at this point of time is Direct Current (DC) and must be transformed in the form of Alternating Current (AC) which is suitable for use in the homes or businesses.

2.5.3. Inverter

Conventionally, an inverter is situated in a location that is accessible, as close as practical to the modules. In a residential application, the inverter is every so often placed to the exterior side-wall of the home neat to the electrical main or sub-panels. As, an inverter is quite noisy, this should be taken into the consideration, at the time of choosing the location of the inverter.

The work of an inverter is to turn the direct current (DC) which is produced with the help of the solar panels into 120 volts AC, that can be put into immediate use with the help of linking the inverter directly to a dedicated circuit breaker in the electrical panel (Figure 2.10).

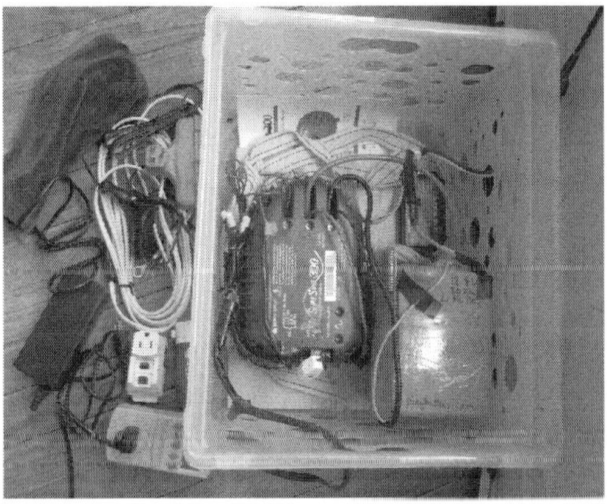

Figure 2.10. An illustration of solar inverter.

Source: Image by Flickr.

The inverter, electricity net meter, along with the electricity production meter is linked, in such a way that the power that is being generated with the help of solar electric system will first be consumed with the help of the electrical loads that are currently in operation.

The balance of the power or electric energy which is generated with the help of the solar electric system passes through the electrical panel and out on to the electric grid.

2.5.4. Net Meter

With respect to the solar electric system that is also connected to the utility grid, the DC power from the solar array is then transformed into 120/240 volts AC power and fed directly into the utility power distribution system of the building.

The power is "net metered." In simple words, the meaning of net-metered is, that it decreases the demand of the power from the utility when the solar array is producing the electricity, in this way it decreases the utility bill.

These grid-tied systems automatically turned off if the utility power goes offline, safeguarding the workers from the power being back-fed into the grid through the course of an outage. These sorts of electric systems that are driven with the help of solar systems are known as "on grid" or "battery-less" and make up near about 98% of the solar power systems being installed in present interval of time.

2.5.5. Other Benefits of Solar

With the help of decreasing the utility bill of a building, these kinds of systems not only pay for themselves over the interval of time, but in addition to this, they also aid in decreasing the air pollution that is being caused by the utility firms.

For instance, solar power systems provide assistance in order to elevate something that is known as "peak load generating capacity," in this way, saving the utility from turning on expensive and polluting supplemental systems through the course of the peak demand.

The more locally produced solar electric power systems that are installed in a given utility's service area, the less capacity the utility require to build, in this way, saving each and everybody from funding costly additional power generating the sources.

Contributing clean, green power with the help of solar electric systems provide assistance in order to generate employment and is a great way to decrease the pollution along with the other issues that are produced by the electricity that is being generated with the help of fossil fuels.

Electricity generating systems that are driven on solar energy provide assistance in order to decrease the effect on the environment and also, it is cost-effective at the same point of time.

2.6. CONSIDERATIONS TO KEEP IN MIND BEFORE INSTALLING SOLAR PANELS

Before taking any step, in order to install a solar panel in the home, ensure that the solar power befits the home. Then make sure that the appropriate amount of light from sun or solar radiation is available at site. The suitability of the solar energy fluctuates significantly with how much sun an area receives. If the area does not receive enough amount of solar radiation or sunlight, putting money in the solar panels will not be the idea to take.

After identifying that the area receives sufficient amount of solar radiation or light from sun, ensure the installation space is available. If an individual has an open space to spare in the backyard, it would be a perfect site for ground mounts. The backyard option is ideal for those roofs are heavily shaded or not structurally essential for the solar panels.

In addition to this, it is very important to acknowledge that the local laws with respect to the solar panel installation in order to avoid getting into the trouble with the local authorities. An individual can get the information from their local solar energy consultant.

2.6.1. Environmental Impact

Nevertheless, the solar energy is well thought out to be one of the best, cleanest as well as renewable source of the energy between the available sources, it has some environmental effect as well. solar energy employs PV cells or solar cells in order to generate the solar power.

Although, manufacturing the PV cells in order to generate that energy necessitates silicon and produce some waste products. Inadequate regulation of these sorts of material can take into the direction of dangerous exposure to humans as well as environment.Installation of the solar power plant may necessitate a huge area of land, which can affect the existing ecosystem. Solar energy does not pollute the air when it is transformed in the form of the electricity with the help of the solar panels. Solar energy found in plenty amount and in addition to this, solar energy does not help in global warming.

2.7. IMPORTANCE OF SOLAR ENERGY

Solar energy is the prime major renewable source. Solar energy is capable in order to fulfil the several numbers of difficulties all across the globe. There are several numbers of various reasons in order to promote its share in the industry of energy. This power source is achieving more and more popularity, and the reason behind this is its versatility with numerous advantages to people as well as the environment.

As per the United States DOE, the amount of solar radiation or light coming from the sun received by the Earth in one hour is more than the total energy used in an entire year, all across the globe.

In the year 2015, solar energy was the fastest energy sector with an increment of near about 33%, as per the reports of Bloomberg. The environmental benefits are the main drivers in order to promote the solar energy.

2.7.1. The Promise of Solar Energy

All life on the Earth is supported with the help of the Sun. This amazing resource radiates huge amount of energy and gives us both heat as well as light with the help of fusing the hydrogen into the helium at it once. This process is known as solar radiation. Only about half of this solar radiation makes it to the surface of the Earth.

The rest of the solar radiation is either absorbed or reflected with the help of the clouds as well as the atmosphere. But still, enough amount of the power that is coming from the sun is received by the Earth, in order to fulfill the requirement of power of all mankind, which is millions of times over. Solar energy, which is the power coming from the Sun, and is vast, never-ending, and a clean source.

Solar radiation, or solar energy can be used directly in order to heat or lightning the homes and businesses, for producing the electricity, and for hot water heating, solar cooling, and a several kinds of other commercial as well as industrial uses.

Most critical, given the increasing concern over the sudden changes in the climate, is the fact that the production of the electricity with the help of the solar radiation represents a clean substitute to the electric energy from the fossil fuels, no global warming pollution, no risks of the electricity price hikes, and no dangers to the human health.The solar resource is massive. As per the United States DOE, the amount of the sunlight or the solar radiation that reaches the surface of the Earth in an hour and a half is sufficient in order to control the consumption of the energy of the entire world for a year. Just 18 days of sunshine on the Earth is consisting of the same amount of the energy as is stored in all of the reserves of natural gas, coal, and oil of the planet.And, once a system is in place in order to capture the solar resource and transform it in the form of useful energy, the fuel is free.

2.7.2. The Growth of Solar Energy

From the year of 2008, the U.S. installations have increased seventeen times from the value of 1.2 gigawatts (GW) to an estimated 30 GW till 2019. It is sufficient amount of power which can power near about 5.7 million average American homes.

From the time period of 2010, the average cost of the SPV panels has decreased over 60% and the cost of a solar electric system has dropped by near about 50%.

Solar power is now well thought out as an economically competitive with the traditional energy sources in the various kinds of states, which is consisting of California, Hawaii, Texas, and Minnesota.

2.7.3. The Basics of Solar Energy

What are the basics of the solar energy systems? First of all, solar energy systems fluctuate according to the size as well as the usage. Residential systems are found on rooftops all across the United States, and the businesses are starting to install the solar panels in order to reduce or lessen their utility bills or energy costs.Utilities are also building large solar power plants in order to provide the cleaner energy to the all customers that are linked to the grid.Irrespective of the specifics of a given installation, there are two main kinds of solar energy technologies, that are photovoltaic (PV) and concentrating solar power (CSP). Most of the people are familiar with the photovoltaic technology, and the reason behind this is that the solar panels they see more and more on the tops of the buildings or ones placed on the International Space Station.

When the sun shines onto one of those solar panels, photons from the solar radiation or the light coming from the sun are absorbed with the help of cells that are present in the panel, which produced an electric field across the layers and causes electric current in order to flow. The installation of the photovoltaics may be ground-mounted, rooftop mounted and wall mounted as well.The photovoltaics may be mounted in a permanent orientation in order to maximize the generation of the electricity and value or they may be mounted on trackers that follows the sun across the sky. Rooftops photovoltaic panels make solar power viable in virtually each and every part of the United States.In a sunny location, for example, Los Angeles or Phoenix, a five kilowatts residential system generates near about of seven thousand kilowatts to eight thousand kilowatts with each passing day, which is approximately equal to the electricity usage of a typical United States household.

The second technology is focusing on the solar power (which is also known as the concentrated solar thermal and CSP). Concentrated solar thermal is mainly used in very large power plants and is not suitable for the residential use.

This technology employs mirrors in order to reflect and focus the solar radiation on to the receivers that gathers the solar energy and then transform it in the form of heat. The heat is then used in order to drive the heat engine, which is generally a steam turbine, that is linked with an electric power generator which is then used to generate the electric energy or electric current.

2.8. ADVANTAGES OF SOLAR ENERGY

2.8.1. Renewable Energy Source

Out of all the advantages of the solar panels, the most necessary thing is that the solar energy is a truly renewable source of the energy. Solar energy can be harnessed in all regions of the world and also it is available for each and every day.

Solar energy is never ending source of energy, it is not like some of the other source of the energy. Solar energy will be accessible as long as Sun exists, and in this way, the solar radiation will be available for every person for at least 5 billion years when according to the scientists the sun is going to die.

2.8.2. Reduces Electricity Bills

Since, people will be fulfilling some of their energy requirement with the help of the electric power that has been produced with the help of the solar system, their utility bills will drop. How much a person save on his bill will rely on the size of the solar system as well as the usage of the heat or the electricity.

In addition to this, not only will a person will be saving on the electricity bill, there is also a probability in order to receive payments for the surplus energy that an individual export back to the grid. If a person produces more electric energy as compared to the use (considering that the solar panel of that person is linked to the grid).

2.8.3. Diverse Applications

The energy produced with the help of the solar system can be used for several kinds of various purpose. A person can produce electricity (with the help of photovoltaics) or they can produce heat (solar thermal). Solar energy can be used in order to generate the electricity in areas without the access to the energy grid, to distill water in regions with limited clean water supplies and in order to power the satellites that are present in the space.

Solar energy can be combined into the materials that are used for buildings. Not long ago Sharp introduced transparent solar energy windows.

2.8.4. Low Maintenance Costs

Generally, solar energy systems do not necessitate a lot of maintenance. A person only requires to keep the solar energy systems relatively clean, in this way, cleaning them a couple of times per year will do the job. If in doubt, a person can always rely on the specialized cleaning organizations.

Most dependable solar panel manufacturers offer a warranty of near about 20 years to 25 years. Also, there are no moving parts, there is no wear and tear. The inverter is only the part, out of all, which requires to be changed after the time period of near about 5 years to ten years, and the reason behind this is that, inverter is working non-stop in order to transform the solar energy in the form of electric energy and heat (SPV versus solar thermal).Apart from the inverter, the cables also require to be maintained in order to make the power system runs at maximum efficiency. In this way, after transforming the initial cost of the solar system, a person can expect very little spending on maintenance and repair work.

2.8.5. Technology Development

Technology in the solar power industry is consistently improving and enhancements will be increased in the coming interval of time. Innovations in quantum physics and nanotechnology can potentially elevate the efficiency of the solar panels and double, or even three times, the electrical input of the solar power systems.

2.9. DISADVANTAGES OF SOLAR ENERGY

2.9.1. Cost

The initial cost of purchasing a solar system is fairly high. This is consisting of paying for the solar panels, inverters, batteries, wiring, as well as for the installation. Nonetheless, the solar technologies are continuously developing, in this way, it is safe to assume that the cost will decrease in the coming interval of time.

2.9.2. Weather Dependent

Nevertheless, the solar energy can still be gathered through the course of cloudy as well as rainy days, the efficiency of the solar system drops. Solar panels rely on the solar radiations or light coming from the Sun in order to efficiently collect the solar energy. In this way, a few cloudy, rainy days can have a considerable impact on the energy system.

A person should also take into consideration that the solar energy cannot be gathered through the course of the night. On the contrary, if a person also needs his or her heating solution to work at night or through the course of the wintertime, thermodynamic panels are an alternative to consider.

2.9.3. Solar Energy Storage Is Expensive

Solar energy has to be used right away, or the solar energy can be stored in the batteries. These batteries, used in off-the-grid solar systems, can be charged through the course of the day, in such a way that the energy can be used in the night. This is good solution for utilizing the energy all day along but in addition to this, it is also quite expensive.

In most of the instances, it is smarter to just use solar energy through the course of the day and take energy from the grid through the course of the night (a person can only do this if the system is linked to the grid). Luckily, the requirement of the energy is generally higher through the course of the day, in this way, a person can fulfill his or her requirement of the energy with the help of the solar energy.

2.9.4. Uses a Lot of Space

The more electricity a person desire to generate, the more numbers of solar panels he will need, as he or she wants to gather much solar radiation as possible. SPV panels necessitates a lot of space and some roofs are not big enough in order to fit the number of solar panels that he or she would like to have.

2.9.5. Associated with Pollution

Never the less, the pollution that is related to the solar energy system is far less when it compared with the other sources of energy, solar energy can be linked with the pollution. The transportation as well as the installation of the solar system have been linked with the emission of the green-house gases (GHG).

2.10. CONCLUSION

Solar power or solar energy is a massive source of directly useable energy and with the help of solar radiations, several numbers of various devices can be drive or powered and can be charged.

Most of the surface of the Earth receive enough amount of solar energy in order to permit low-grade heating of water and buildings, never the less, there are huge amount of variations with the latitude and season. At low latitudes, simple mirror devices can focus solar energy sufficiently for cooking and also, in order to power the steam turbines.

The energy of light shifts electrons in some of the semiconducting materials. This photovoltaic effect is having the potential of producing the electricity in large scale. Never the less, the present low efficiency of the SPV cells requires very large areas in order to supply the electricity demands.

Direct use of the solar energy or the solar power is the only renewable means that is having the potential of ultimately supplanting existing global energy supply from nonrenewable sources, but at the expense of a land area of at least half a million kilometers.

REFERENCES

1. Ashok, S., (2020). Solar Energy | Description, Uses, and Facts. [online] Encyclopedia Britannica. Available at: https://www.britannica.com/science/solar-energy#ref339542 (accessed on 20 August 2020).
2. Dana, O., (2019). Why the Future of Solar Is Bright -Renewable Energy World. [online] Renewable Energy World. Available at: https://www.renewableenergyworld.com/2019/11/12/why-the-future-of-solar-is-bright/#gref (accessed on 20 August 2020).
3. Freedom solar power, (2018). 7 Uses of Solar Energy | Freedom Solar Power. [online] Freedomsolarpower.com. Available at: https://freedomsolarpower.com/blog/7-uses-of-solar-energy (accessed on 20 August 2020).
4. Green match, (2020). Advantages and Disadvantages of Solar Energy (2020) | Green match. [online] Greenmatch.co.uk. Available at: https://www.greenmatch.co.uk/blog/2014/08/5-advantages-and-5-disadvantages-of-solar-energy (accessed on 20 August 2020).
5. Marsh, J., (2020). What is Solar Energy? Solar Power Explained | EnergySage. [online] Solar News. Available at: https://news.energysage.com/what-is-solar-energy/ (accessed on 20 August 2020).
6. PV education, (2019). Solar Energy | PV education. [online] Pveducation.org. Available at: https://www.pveducation.org/pvcdrom/introduction/solar-energy (accessed on 20 August 2020).
7. Richardson, L., (2018). The 5 Most Common Uses of Solar Energy in 2020 | EnergySage. [online] Solar News. Available at: https://news.energysage.com/most-common-solar-energy-uses/ (accessed on 20 August 2020).
8. Rinkesh, (2020). What is Solar Energy and How Solar Energy Is Converted into Electricity-Conserve Energy Future? [online] Conserve Energy Future. Available at: https://www.conserve-energy-future.com/solarenergy.php (accessed on 20 August 2020).
9. Solar Electric Power Company, (2012). Solar Power Advantages and Disadvantages. [online] Sepco-solarlighting.com. Available at: https://www.sepco-solarlighting.com/blog/bid/115086/solar-power-advantages-and-disadvantages (accessed on 20 August 2020).
10. Solar, N., (2015). How Do Solar Systems Produce Energy? [online] NW Wind & Solar. Available at: https://www.nwwindandsolar.com/solar-power-in-seattle-and-the-northwest/how-do-solar-systems-

produce-energy/ (accessed on 20 August 2020).

11. Venture Solar, (2017). The Importance of Solar Energy | Venture Solar. [online] Venture Solar. Available at: https://venturesolar.com/blog/the-importance-of-solar-energy/ (accessed on 20 August 2020).
12. Vijayalaxmi, K., (2020). Why is Solar Energy Important? [online] Greenliving.lovetoknow.com. Available at: https://greenliving.lovetoknow.com/Why_Is_Solar_Energy_Important (accessed on 20 August 2020).

Chapter

3 Benefits and Risks of Solar-Powered Irrigation

CONTENTS

The chapter of benefits and the risks of the solar-powered irrigation explains the increasing interest in the solar-powered irrigation system. This chapter also explains the working of the solar irrigation system and where is solar irrigation takes place. This chapter also emphasizes the various components of the solar-powered irrigation system such as motor pump, pump controller, and solar panel. This chapter mentions the various kinds of the motor pumps such as surface pump and submersible pump.

This chapter explains the basic significance of the automatic solar submersible pump control for irrigation along with the components of the automatic solar submersible pump control such as the submersible pump, PV panels, charge controller, battery, inverter, and inverter. This chapter also explains the working of the solar-powered auto irrigation system.

This chapter also addresses the approximated cost in order to install and run a solar-powered irrigation system. This chapter also provide highlights on the factors that are need to consider when purchasing a solar power irrigation system such as quality, durability, price, efficiency, warranty, and warranty of the equipment.

This chapter also explains the solar-powered irrigation system toolbox along with its operating principle, components of the solar-powered irrigation toolbox, and the configuration of the solar-powered irrigation system. This chapter also describes the advantages as well as the disadvantages of the solar power irrigation systems.

3.1. INTRODUCTION

All across the globe, agriculture is the single largest employer. Agriculture sustains the livelihoods of near about 40% of the population all across the word, out of which who continue to live in poverty (United Nations, 2015). Irrigation is between the measures that can enhance yields, decrease the vulnerability in order to change rainfall patterns and permit several kinds of cropping practices (FAO, 2011).In this way, the irrigation is every so often seen as the engine that provide assistance in order to make sure the security of food, produces incomes, offers jobs and drives rural development. Energy is the main input for irrigation services.

As the financial inputs for the solar-powered irrigation systems (SPIS) are decreasing, and subsidy schemes for the solar-powered input systems are being rolled out, the solar techniques are becoming an important option for both large-scale farmers as well as small-scale farmers.

Solar power irrigation system offers a reliable as well as cost effective energy, potentially lesser cost of energy for the process of irrigation. With respect to the rural regions, where the diesel fuel is very costly or where the reliable access to the electricity grid is absent, they can offer a comparatively flexible and climate friendly substitute source of energy. Solar-powered irrigation system can be employed in large-scale irrigation systems as well as for decentralized, small-scale irrigation.

In addition to this, several nations are encouraging the solar-powered irrigation system in the framework of national action plans with respect to the sudden changes in the climate as a way, in order to decrease the emission from the agriculture. The operation of the solar pumps does not generate any kind of emission of green-house gas (GHG).

Life cycle assessments of the solar power irrigation systems is responsible for the emission in a cradle-to-grave scenario, which shows a potential decrement in the emission of the GHS per unit of energy which is being used for the water pumping (CO_2-eq/kWh) of 95% to 97% when it is compared with the pumps that operated with the grid electricity (global average energy mix) and 97% to 98%, when they are compared with the diesel pumps (GIZ, 2016).

None the less, it is very important in order to note that the solar-powered irrigation system, if not sufficiently regulated and managed, it cannot bear the risk of the supporting unsustainable water use. Once the installation process of the systems is done then there is no cost per unit of power and, in this way, no financial profit for farmers in order to save on fuel or electricity for water pumping.

This can take into the direction towards the wasteful water use, over-abstraction of groundwater, and low field application efficiency. In some instances, farmers sell water to their neighbors at a profit, increment in the overall water withdrawals.

Identifying the risks that are related to the water and coping up with them from the very starting, more particularly in the financing as well as design stages. This will be very crucial or important in order to make sure the sustainable application of the solar-powered irrigation system technology.

In the context of the quick expansion or rapid development of the solar-powered irrigation system, there is a chance to not simply introduce a clean, climate-smart, as well as an innovative energy technology, but to think in a strategic way regarding how this technology can be employed in order to promote more sustainable use of groundwater resources, in order to produce more and more inclusive finance as well as management structure and to prosper more combined thinking about the solutions around the water-energy-food nexus.

3.2. INCREASING INTEREST IN SOLAR-POWERED IRRIGATION SYSTEM

There is an increment in the interest in the solar power irrigation systems all across the globe, noticeable in more and more frequent requests from the agricultural institutions in the emerging nations for the installation, finance, and training.

In the year of 2015, the Food and Agriculture Organization of the United Nations (FAO) and the Deutsche Gesellschaft fürInternationaleZusammenarbeit (GIZ) GmbH have hosted an exploratory workshop in order to better understand the potential of the solar powered irrigation system for emerging nations.

Representatives from near about nineteen nations shared their experience with the solar pumping technologies, from large scale to small scale, from tropical climate zones to arid climate zones, for vegetable gardens, orchards, and livestock watering, with the help of the surface and groundwater.

It became very clear that there was a requirement in order to synthesize these kinds of experience and to respond to the problems, questions, and needs that were arose through the course of the workshop. What is the actual cost of the of solar power irrigation system, and what are the advantages of the solar power irrigation system as compared to the other technologies?

What kinds of policies, rules, and regulations are required in order to handle or cope up with the risks and realize the potential of the solar power irrigation system? What are viable business models for the solar power irrigation system? How can small-holders gets the advantages from the technology of solar power irrigation system? How can the risk of the groundwater depletion be addressed efficiently?

How can solar power irrigation system provide assistance in order to empower women and encourage the gender equity? What kinds of capacity development programs are required in order to support the farmers, extension workers, local private sectors and others? What are the chances for knowledge exchange and technology transfer? (Figure 3.1).

Figure 3.1. Solar panel energy panels installed in the field.

Source: Image by Pixabay.

Food and Agriculture Organization of the United Nations (FAO) and the Deutsche Gesellschaft fürInternationaleZusammenarbeit (GIZ) decided to follow up on these kinds of queries with a global project that is the part of the "Powering Agriculture: An Energy Grand Challenge for Development" (PAEGC) initiative. The main objective of the project is to:

- i Learn from good practices all across the world;
- ii Foster policy dialogue across sectors on topics for example, finance, green jobs and groundwater management; and
- iii Provide assistance to the responsible use of solar power irrigation system with the help of the training and enhanced advisory services.

3.3. SOLAR-POWERED IRRIGATION: A SOLUTION TO WATER MANAGEMENT IN AGRICULTURE

Farmer have always played an important role in the society, and the reason behind this is that, farmers provides food to each and every person all across the globe. Never the less, on may forget that, not only the farmers provide food, but in addition to this, the farmers also provide energy, which, in the present interval of time, is of prime importance, more particularly with the respect to the renewable energies.

Certainly, farmers can produce energy with the help of the wind, or with the help of the sun, or with the help of the biomass and they can use it for their own farm, or, if they have a surplus, resell it to the organizations or firms.

Solar energy can be the one of the easiest technique or method for the farmers in order to produce energy. Certainly, farmers generally have various kinds of large buildings whose roofs are directly under the sun or there is adequate amount of solar radiation is present, without being hampered by the shadows of trees, turning them in the form of an ideal place to install a photovoltaic (PV) system.

In this way, the utilization of the solar energy in agriculture is becoming more and more common and the energy generated with the help of this renewable source can be used either on the farm or in the local power grid as well, providing the farmer with an additional incentive or additional income.

One of the fields in the agriculture that provides advantages the most with the help of solar energy is irrigation, more particularly in dry or parched areas. The main reason behind this is that, with the help of the solar radiations or sun for the purpose of irrigation shows a virtuous circle: when the sun shines, it feed the irrigation system, well, it has been known that crops require more amount of water as compared to the amount of the heat. In this way, a huge amount of energy is available when it is actually required.

3.3.1. How Does a Solar Irrigation System Work?

In order to transport the water, the pumps are used which is equipped with the solar cells. The solar energy absorbed with the help of the cells is then transformed in the form of electrical energy or electric current with the help of a generator which then feeds an electric motor which is used to drive the motor.

Most of the conventional pump systems primarily work on a diesel engine or with the local power grid. Never the less, the two modes of the operations shows disadvantages when they are compared to the solar pumps.

In several numbers of various rural regions, more particularly in developing nations, the access to the electricity grid is not always guaranteed. In this case, farmers cannot depend on the conventional irrigation system (Figure 3.2).

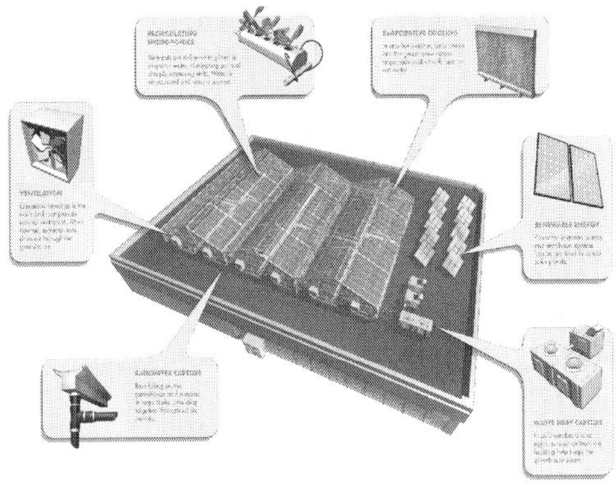

Figure 3.2. An illustration of bright farm system design.

Source: Image by Wikimedia commons.

In this way, with the help of an independent and alternative system for providing energy can be a solution with respect to the farmers in order to safeguard a safe power source and for the public grid in order to avoid saturation.

Diesel pumps are quite more effective, when they are compared with the pumps that works on alternative current. And the reason behind this is that, diesel pumps permit greater flexibility. Never the less, one of the main restrictions is that this system depends on the availability of the fuel, added to a greater effect on the environment.

Pumps that driven with the help of diesel are much cheaper when they are compared with the pumps that are driven with the help of solar energy, but the operating costs are quite high and rely heavily on the price of diesel.

With respect to the solar-powered systems, they work the other way round, that is, never the less this system is comparatively costly, the source of energy is free, in this way, after the amortization period, there are no longer operating costs (only the maintenance costs must be well thought out).

In this way, the solar pumps are turn out be a viable investment for long-term. According to the various studies, for example, Water for wealth and food security by Ag- Water Solutions Project, have depicted, the access to the water for the agricultural purposes remains critical in some regions for example in dry or parched areas of Africa and Southern Asia.

Several numbers of farmers from India and Africa fetch the water directly from the well or the rivers and then irrigate the fields with the help of the buckets. If the farmers of those regions could have access to a motorized pump, they would elevate their harvesting by near about 300%.

In this way, in the present interval of time, Research, and Development tends to concentrate on producing solar pumps that are cost effective in parched or dry regions.

The company named as IBC SOLAR has developed with Siemens, which is a solution in order to substitute the diesel engine with the help of solar-powered engine.

In this case, the entire irrigation system which is also consisting of the pump can remain as it is; only the diesel engine is substituted with the help of a photovoltaic system and the so-called "IBC pump drive controller."

A prototype of the system was tested in the year of 2015 in a farm in Namibia and as per the manufacture, these systems are turned out to be quite effective and cost efficient.

The prime benefit lies in the fact that there are low acquisition costs as the existing infrastructure is used.

3.3.2. Where Is Solar Irrigation Happening?

The installation of the solar pumps in the dry or parched areas for example, Africa, India, and South America is also part of several numbers of development projects, having the objecting in elevating the productivity of the local farmers and as an outcome, enhancement in their living conditions.

One of the successful instances of this is the initiative of a Physics teacher in a school in Blankenese (Germany) where students have created two pumping systems that works on the solar energy in cooperation with the company named as SET GmbH from Wedel (Figure 3.3).

They have installed these solar-powered systems in two farms in Nicaragua in order to pump the underground water. This project could also be accomplished with the help of the UNAN University in León, which deals a lot with the use of solar energy.

Figure 3.3. An illustration or arid soil.

Source: Image by Peakpx.

Certainly, the project has been running for more than ten years and near about thirty solar-powered pumps are in operation now in Nicaragua. The employment of the solar pumps permits the latter in order to generate all across the year, even in dry season and in this way, in order to elevate their income and support their position in the local market.

Apart from the regions that are mentioned in the above topics, in addition to this, there is increment in the interest in the solar irrigation system in Europe. Just a few month ago, a mobile solar drip irrigation system from Austria has reached in the stage of production.

The Austrian company Wien Energie carried out this project which follows a dual objective. Out of which, the first one is the decrement in the emission of the carbon dioxide (CO_2), owing to the use of the solar energy. And the second is, accomplishment of near about 30% of water savings tanks to the drip irrigation techniques versus the conventional sprinkler irrigation.

The basic fundamental of the drip irrigation technique is pretty simple. With the help of the several valves, hoes, and pipes, the water drips slowly and at a frequent period of time to the roots of the plants.

In this way, there is no water waste as water goes directly where it should go, unlike to the sprinkler system in which the water evaporates into the air or seeps into the soils where no plant grows. In this manner, the drip irrigation techniques permit in order to harvest more crops with the help of less amount of water, turning it into a highly effective irrigation method.

With respect to the Wien Energie solar irrigation system, a mobile solar energy system that are equipped with the help of PV modules (up to 3 kW) is linked to a wheeled pump which can pump from wells or rivers.

The pumps that are driven with the help of the solar energy then distributes the water directly to the crops with the help of hoses. After a successful test on a 3.5 hectares organic cornfield in Guntramsdorf, Austria, this system is now ready for production.

In this way, in the nations that are suffering from high temperatures, amid regions, and scarce water resources, the drip irrigation system could contribute to an effective water management. This is all the more important as farmers have to face three difficulties, that are save water, save money, and save energy.

Mobile solar drip irrigation systems shall turn out to be a perfect solution in order to cope up with these kinds of difficulties. Never the less, these kinds of systems are still quite costly and complexed to settle, several numbers of various research and development projects are working on the democratization of the use of the solar power in the field of agriculture, which, in the future (and in present interval of time as well), could play an important role in the management of the food and energy crisis.

3.4. COMPONENTS OF THE SOLAR-POWERED IRRIGATION SYSTEM

The actual components and the hardware requirement relies on the kind of the irrigation system, for example, surface irrigation where the water is moved across the surface of the agricultural lands, localized irrigation such as spray or drip or trickle system where the water is applied to each and every plant or adjacent to is, and sprinkler irrigation where the water is piped to one central location or more than one central locations with in the field and distributed with the help of an overhead high pressure sprinklers or guns.

The main components of a common solar-powered pumping system are consisting of a solar panel array that helps in proving the power to the borewell pump or a surface pump. The size of the entire system is relying on the size of the pump, the amount of water needed per day and the power available from the solar panels (Figure 3.4).

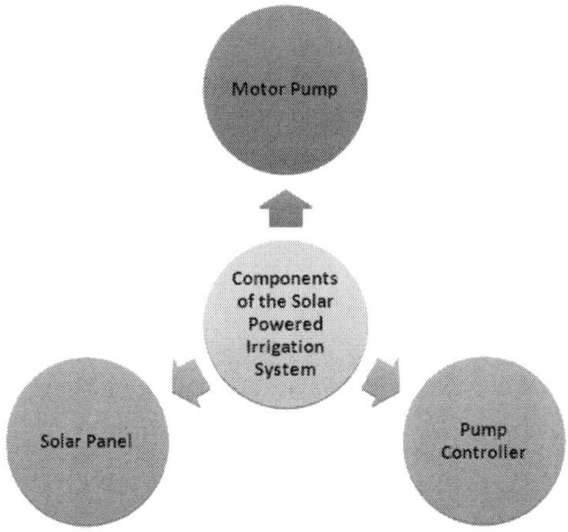

Figure 3.4. Components of the solar-powered irrigation system.

The water is every so often pumped from a bore well or stream into a storage tank or directly into the field. The solar power irrigation system has three main components, that are: Motor pump, pump controller and solar panel.

3.4.1. Motor Pump

The pump is consisting of a motor that run on the electric power that is produced with the help of the solar panel. According to the type of the motor (that means whether it is alternating current (AC) motor or direct current (DC) motor), the voltage of the solar pump motor can be AC or AC.

Direct current motors are mostly used for small applications to the medium application, for example, garden fountains, landscaping, drinking-water for livestock, or small irrigation projects.

Most of the DC pumps run at 24 volts in spite of 12 colts. Never the less, smaller systems for 12 volts do exist for light duty applications. On the other hand, the solar power irrigation systems that are running on the 48 volts also exists for much heavier applications. The power of a motor is specified in wither watts or horsepower (HP).

Table 3.1. Typical AC Pump and Controller Specifications

AC Pump HP	Volts	Controller	Solar Panel (watt-peak)	Battery Capacity
1HP	60	4 kVA	1500 Wp	100 Ah – 5 Nos.

1.5 HP	72	5 kVA	1800 Wp	100 Ap – 6 Nos.
2 HP	96	6 kVA	2400 Wp	100 Ah – 8 Nos.
2.5 HP	96	6 kVA	2400 Wp	120 Ah – 8 Nos.

Source: Table from www.solarenergyproducts.co.in

Table 3.2. DC Pump with Charge Controllers

DC Pump	**0.125 HP**	**0.25 HP**	**0.5 HP**	**1 HP**	**2 HP**
Volts	24 V	24 V	24 V	48 V	48 V
NO- Load Current	2 A	4 A	8 A	8 A	16 A
Full-load Current	4 A	8 A	15 A	16 A	32 A
Suction	1.83 m (6 Feet)	7.62 m (25 Feet)	16.46 m (54 Feet)	76.2 m (250 Feet)	152.4 m (500 Feet)
Delivery of Flow Rate (liters per hour or minute)	500-lph	8-lph	24- lph	35- lph	70- lph
Panel Capacity	200 Wp	500 Wp	800 Wp	1500 Wp	3000 Wp
Battery Capacity	50 Ah–2 Nos.	150 Ah –2 Nos.	200 Ah–2 Nos.	200 Ah–4 Nos.	200 Ah–8 Nos.
Charge Controller Rating	24 V, 40 A	24 V, 40 A	24 V, 40 A	48 V, 40 A	48 V, 40 A–2 Nos.

Source: Table from www.solarenergyproducts.co.in

3.4.2. Pump Controller

There are two kinds of pumps controllers, the first one is the inverter and the second is variable frequency drive (VFD). If an AC solar pump is used, an inverter becomes very significant in order to alter the DC in the form of AC. The supported power range of a common inverter extends from 015kW to 55kW, with the high-powered inverter being used for larger irrigation systems. The solar panel and the inverter must be sized consequently in order to accommodate the inrush characteristics of an AC motor. As the pumps that work with help of AC necessitates high power at the start, the inverter should be able to regulate this extra start-up load. Every so often, a VFD controller is employed in order to make sure the motor of the pump gets sufficient amount of voltage and current. Some typical specifications for several kinds of AC pumps and the pump controllers are listed in the above-mentioned table 3.1.

Several numbers of solar DC pumps necessitate a special type of controller if they are to be driven directly with the help of the photovoltaic modules (or without the help of the batteries). The controller or linear current booster (maximum power point tracker) allows the pumps to start and run in low light conditions through the course of a rainy day, cloudy day, or early morning and evening. With the help of a battery source, the controller may not be required at all for a DC pump. A special kind of controller can also be used in order to transform the 12 volts battery power in the form of higher voltage in order to drive the pump at top speed. The DC pumps are available from 12 V to 48 V with their initial torque fluctuating from 12 A to 96 A. A common specification of different DC pumps and charge controllers that are available from a vendor that are listed in the table 3.2. in the above section as an example.

3.4.3. Solar Panel

The pumps are fitted with the motors, which receives energy from the solar photovoltaic (SPV) arrays. The nominal power of a solar module is expressed in peak wattage (Wp). The wattage of the solar panels relies on the necessity and the usage of the motor.Talking about India, SPV arrays that are having capacity in the range of near about 200 W to 5kWp are suggested by the government. A person may refer to the information with respect to the SPV array and motor pump sets for solar PV water pumping system (2013–2014) given by the Ministry of New and Renewable Energy, under Jawaharlal Nehru National Solar Mission. Several numbers of various vendors offer installation service at a cost. But if a person is planning to install by himself, he or she can refer to several numbers of live video clips that are available on the internet.

3.5. TYPES OF MOTOR PUMPS

Solar pumps fall into two major groups, out of which, first is surface pump and second is submersible pump, according to the construction and applications of them (Figure 3.5).

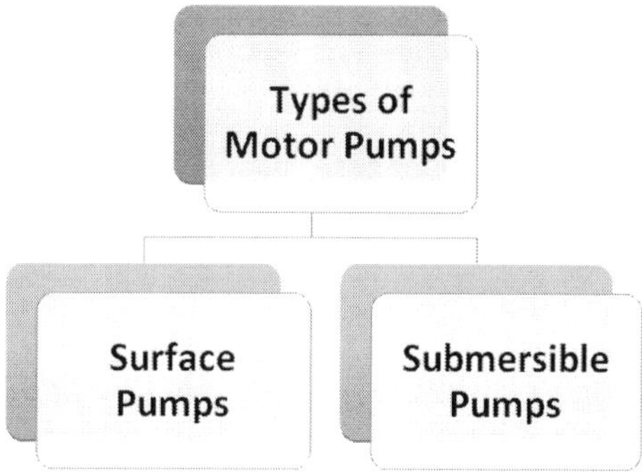

Figure 3.5. Types of motor pumps.

3.5.1. Surface Pump

Generally, the pump is installed above the water level which necessitate a suction lift. Surface pump is used in order to get discharge from an open well, pond or river where high flow and small head (height to which the water is required to be lifted to) are needed.

There are several numbers of various kinds of pumps that are available, and DC surface pumps with lift up to 21m (70feet) are easily available in the India.

3.5.2. Submersible Pump

Basically, the submersible pump is a hermetically sealed motor. The entire assembly of submersible pump is submerged inside the water in order to pump it up. Submersible pump is generally employed for pumping the water from the small-bore wells, and is designed in order to fit inside the drilled bore. The well may be anywhere from a few meters to near about five hundred and fifty meters deep. But the deeper the well, the more powerful and expensive would be the system. The 0.25 HP to 3 HP pumps is easily available in the market. Direct current submersible pumps with total up to near about 92 meters are quite common in India.

3.6. AUTOMATIC SOLAR SUBMERSIBLE PUMP CONTROL FOR IRRIGATION

These systems work under the sun-light. When sun shines the water pumping process is a sensible way of solar electric power utilization all across the course of summer, as the requirement of the water is the highest. These kinds of pumps provide a dependable and consistent source for the plantation.

With respect to any kind of pumping system which is based on the solar power, the potential in order to drive water is a function of three variable for example, power, flow, and pressure. There are several numbers of components are used in this automatic solar submersible pump control panel for irrigation, out of which the main or important components are mentioned as follows.

3.6.1. The Submersible Pump

This solar system is consisting of a submersible pump, an irrigation pump, a water reservoir, and linked water pumps. In the site trail, the submersible pump is kept in a case which is made up with the help of stainless steel, and this case is placed in a well pit at the junction of the open channel and the natural stream course (Figure 3.6).

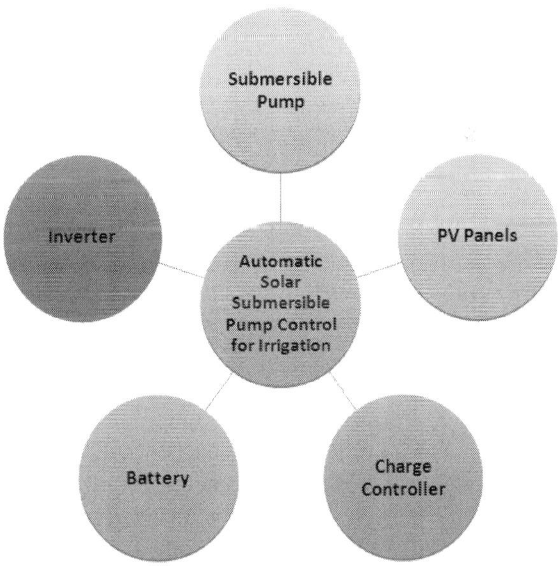

Figure 3.6. Automatic solar submersible pump control for irrigation.

The pump controller pumps the water to the tank at a particular interval of time as set in the control unit. This system is designed with 450 W of power, which is capable of pumping around 2000 liters of the water within 60 minutes or in an hour. This power capacity takes into the consideration the height difference between the water tank and the solar submersible pump.

3.6.2. The PV Panels

The photovoltaic cells rely on the size of the pump. A panel is esteemed in watts of power it can produce. This solar submersible pumping system should be operated with a photovoltaic array of capacity in the range of two hundred watts to five hundred watts peak, and measured under some standard test conditions.

A plenty of modules in series and parallel could be used in order to get the necessary photovoltaic array power output. The output of the photovoltaic modules which are used in the photovoltaic array standard test conditions should be a main of seventy-four watts peak.

3.6.3. Charge Controller

A solar charge controller plays a very significant part in any kind of solar power system. charge controller is used in order to maintain the sufficient charging voltages of the batteries. The charges controller controls current and voltage from the solar panel and charges the battery, and also stops the charging of the battery from over and undercharging settings.

3.6.4. Battery

The battery is an electric device. Battery is used in order to store the electric energy which is generated with the help of the solar panel and supplied to the corresponding loads. The number of batteries that are needed relies on the requirement of the load.

3.6.5. Inverter

The prime function of the inverter is that it transforms the voltage of the battery in the form of AC voltage in order to activate then loads. In this way, it provides assistance in order to run several numbers of electronic devices, home appliances and computer. In the present interval of time, there are several numbers of various kinds of inverters that are available in the market.

The properties of a conventional inverter are consisting of a high switching frequency, high conversion frequency and less harmonic content, etc.

An instance project for the above automatic water pump controller plant irrigation system with is the Solar-Powered Auto Irrigation System. The description of this project is mentioned below.

3.6.6. Solar-Powered Auto Irrigation System

The prime objective of this project is to create an irrigation system in the field of the agriculture with the help of the solar energy and it has several numbers of benefits as well.

The needed components are 8051 series Microcontroller, 12V DC mini submersible pump, Op-Amp, LCD, Solar Panel, MOSFET, Relay, Motor, Voltage Regulator, Diodes, Capacitors, Resistors, LED, Crystal, and Transistors

The power supply is made up with the help of step-down transformer, voltage regulator. In which the step-down transformer provides assistance in order to step down the voltage to volts AC, and a bridge rectifier which helps in transforming the AC in the form of DC, then a voltage regulator controls the voltage to 5V which is then employed for the microcontroller operation.

Talking about the solar-powered auto irrigation system project, the solar energy is used in order to activate the irrigation pump. In addition to this, the solar-powered auto irrigation system project is consisting of several sensor parts, which are assembled with the help of an op-amp IC (operational amplifier IC). Op-amps are designed as a comparator. Two coppers wires are injected into the soil in order to sense the situation of the soil, whether the soil is wet or dry.A microcontroller in this project is employed in order to regulate the entire system with the help of noticing the sensors. When the sensors sense the condition of the soil as dry, then with the help of comparator, the commands are being sent to the microcontroller, and also it sends instruction to the relay-driver IC then, it reminds the motor to the pump water to the crops.

In this setting, the comparator plays a role of an interface between the sensing arrangement and the microcontroller the status of the soil, and the water pump is displayed on the LCD (which is short form for Liquid crystal display), which is interfaced to the microcontroller. In the similar manner, when the sensor senses the condition of the soil as wet, then the microcontroller sends the instruction to the relay in order to switch off the motor.

3.7. HOW MUCH IT COSTS TO RUN A SOLAR-POWERED IRRIGATION SYSTEM?

The cost relies on the factors that are mentioned below:

1 Price and number of KWP solar panels needed in order to cater to the energy requirement;
2. Depth of the water source;
3. Water discharge requirement or the type of soil; and
4 Type of motor pump needed (surface or submersible).

The cost of photovoltaic panels has intensely reduced through the course of the last ten years and it is anticipated to have a more considerable drop in prices in the year of 2025.Apart from this, the other components for example pumps and configurations have also reduced.

Cost approximation for Solar-Powered Irrigation System in the time period of mid-2017 for Small-scale complete system in Kenya with a submersible pump, 300 W panels on secured 3 m high stand with controller, filters, and 1-acre drip irrigation, with planning, installation, and guarantees for the pump, panels, and drip lines starts from 2,400 US dollars.

3.8. WHAT TO CONSIDER WHEN BUYING SOLAR PANELS FOR THE SPIS?

Through the course of opting the quality solar panels for the irrigation system, a person or an individual should well thought out about the following criteria that are mentioned below:

3.8.1. Quality

1. ISO certified panels indicate quality; and
2. The International Organization for Standardization (ISO) 9000 series assures the standards of the quality for the manufacturing industry.

3.8.2. Durability

1. Panels should fulfill the International Electro-technical Commission's (IEC) 61215 reliability standard to be considered durable; and
2. Solar panels are engineered in order to survive in extreme setting such as harsh winters and strong winds.

3.8.3. Price

The cost of solar panels is equal to the price per watts plus installation and permits;

It is not guaranteed and essential that the most expensive panels are the best ones; and

The average solar panel cost in 2020 is near about $ 2.99/watt (Figure 3.7).

Figure 3.7. Things to consider when purchasing solar panels for solar-powered irrigation systems.

3.8.4. Efficiency

- Solar panels can transform the solar radiation in the form of usable electric energy.
- The most effective solar panels of 2020 are the Sun- Power X-Series Solar Panels with a record-breaking of 22.80%.
- Higher efficiency rating is better, and the reason behind this is that, it means that more electricity is transformed with the help of the solar radiation in order to generate up to 75% more energy per square foot for over the time period of first 25 years.
- The top solar panel manufacturers by efficiency are Sun Power, LG, REC Group, Panasonic, and Solar tech Universal.

3.8.5. Warranty

As the solar panels is a long-term and costly investment, it is very necessary to know that Solar panels have two (2) warranties, as mentioned below:

1. Performance warranty-Solar panels degrade over the interval of time just like any other product; and

2. The standard performance warranty of the industry is around 25 years.

Performance warranty makes sure that the product will perform and generate electric energy at a certain level of efficiency after a particular number of years.

3.8.6. Equipment Warranty

1. It covers any physical faults or issues that will rise from the manufacturing errors for example glass imperfections, damaged cell connections, loose junction boxes, and so on;
2. A standard product warranty is near about ten years to twelve years;
3. Some products offer 25 years of warranty such as Sun Power, Solaria, LG, Panasonic, and Silfab; and
4. It is better to consult or work with a solar energy consultant before purchasing and installing the solar panels, and the reason behind this is that, they are professionals and they know much more about this industry.

3.9. SOLAR-POWERED IRRIGATION SYSTEM TOOL-BOX

With the help of the solar energy for irrigation makes a lot of sense. Out of which, the first is, irrigation is every so often in rural regions with poor access to the reliable electric energy or supplies of fossil fuel. The second reason is, solar radiation or sunlight is a never-ending resource, more particularly in the areas where rain water scarcity makes irrigation important in the context of the food industry and international trade (Figure 3.8).

In the end, Solar-Powered Irrigation Systems (SPIS) passively self-regulate, due to the volume of the water pumped elevates on a clear hot day when plants necessitate more amount of water, and vice-versa. It is very necessary in order note that a SPIS is more than just a solar pump that are used for irrigation purpose.

Panels, pumps, and irrigation systems are designed on the basis of the availability of the water and requirements of the local crop water. SPIS is a system where the several numbers of various components, from pump to plant, are combined and harmonized.

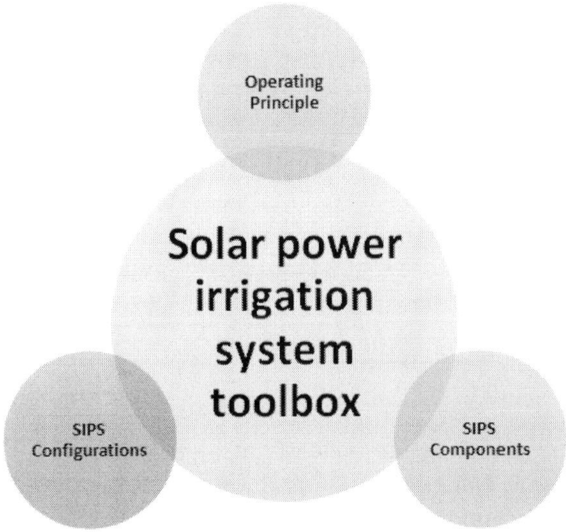

Figure 3.8. Solar power irrigation system toolbox.

3.9.1. Operating Principle

The operating principle of a SPIS is simple. A solar generator gives electric energy for an electric motor pump, which transports the water either directly into an irrigation system or to an increase the level of water in the reservoir. Basic design criteria for SPIS is consisting of a minimum maintenance, maximum reliability, along with the resource efficiency.

A specific aspect of SPIS is the fact that generally a battery back-up is not needed. This is a benefit, and the reason behind this is, batteries are maintenance-intensive, expensive in nature and necessitate frequent replacement.

3.9.2. Components of Solar-Powered Irrigation Systems

The individual components of a SPIS are introduced in the following sections. Table 3.3 shows that, according to site-specific conditions and the potential of the farmer, several kinds of technology options are available. The components along with the alternative technologies can be integrated with each other in an extensive range of ways, but some configurations are better according to the situation in the field.

Table 3.3. Major Alternative Technologies for the Most Important Components of an SPIS

Component	Alternative technologies		Dependent on:
Solar system	fixed	tracking	costs and maintenance intensity
Pump	Surface	Submersible	costs and (geo-)hydrology
Reservoir	reservoir	no reservoir	costs and irrigation system
Irrigation system	Surface	drip or sprinkler	costs and pumping system

Source: Table from energypedia.info.

3.9.3. Solar Power Irrigation System Configurations

The most common solar-powered irrigation system configuration is when a solar generator on a fixed mounting structure offers electricity for a submersible pump which is installed in the borehole. The water is then pumped into the reservoir which is elevated a few meters above the field where it is stored at a constant pressure.When it is released, it flows into a drip irrigation system, which works on the low pressure, and where the water is being filtered and being mixed with the fertilizer before reaching the plants.Never the less, the installation of the water filter on the exit or output side of the tank can be a cause of problem, because the pressure losses in the filter can easily reach several meters, and then, at the low tank heights, there is no water flowing any longer. In this way, it is recommended in order to install the filter on the input side, in order to keep clean water in the tank.

This configuration would also be possible with the help of a tracking system, but would necessitate a higher financial input and more maintenance as compared to a fixed mounting of the solar panels. The reservoir offers stable pressure and water supplies to the drip irrigation system in order to make water distribution as uniform as possible.

The performance of the drip irrigation reduces as the drippers get clogged because of the small particles that are present in the water. Filters prevent this, but only when properly designed for the specific kind of quality of water and irrigation system, and only when is cleaned frequently.

In this way, it is strongly referred to employ the drip irrigation only in the combination with the groundwater as the groundwater is much cleaner as compared to the water which comes from the rivers or open reservoirs. In addition to this, it is suggested to have a monitoring system installed between the pump and the reservoir in order to measure the flow of water and the pressure of the water.

The simplest solar power irrigation system configuration is when a solar generator on a fixed mounting structure that offers electricity for a surface pump which is installed at the reservoir or river. The water is then pumped directly into the surface of irrigation system, for example, with the help of a network of open canals. With respect to this configuration, the pumped water does not pass through an elevated reservoir.

The pressure of the water as well as the flow of the water to the irrigation system corresponds to the actual solar irradiance, which fluctuates over the course of the day, more particularly, with a fixed mounted solar generator. The main benefit of this configuration is the simple installation and comparatively low costs.

Never the less, the main limitation of this configuration is that the producer has little command on the distribution of the water in the field over the days, because there is no reservoir that controls the flow of the water as well as the pressure of the water. The producer will have to use volumetric valves or separate his field into manageable sub-sections in order to regulate the watering of the crops as best as possible.

3.10. ADVANTAGES OF SOLAR-POWERED IRRIGATION

Most recently, several numbers of farmers are turning towards the use of the solar energy for all their irrigation requirements, and it is so with good reason. Farming can be quite expensive endeavor, more particularly, when done on a large scale (Figure 3.9).

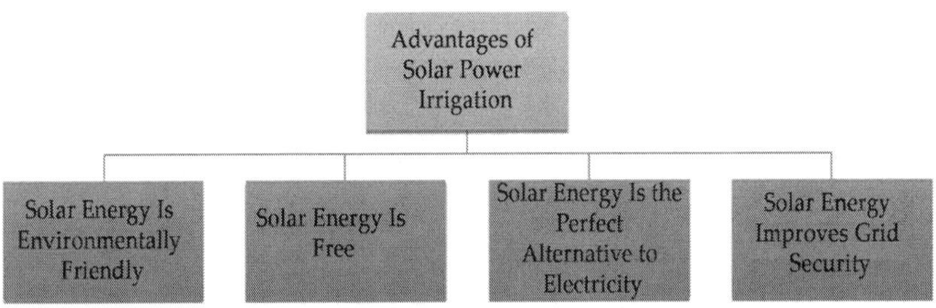

Figure 3.9. Advantages of the solar power irrigation.

In this way, the incorporation of the solar irrigation goes a long way in unbalancing some of those over-the-head costs, which can, as an outcome, be channeled to the other more critical sectors of the farm. There are various numbers of reasons that a person is required to have solar irrigation installed on the farm:

3.10.1. Solar Energy Is Environmentally Friendly

In spite of making sure that the crops are well watered each and every day, solar energy is also known to be environmentally friendly. Not similar to the several energy sources for example, biogas, and coal, solar power does not emit greenhouse gases. As a person already know, greenhouse gases have wreaked some critical chaos on the ozone layer for the past several numbers of years.

Solar irrigation is also ideal, and the reason behind this is that, it does not produce noise as its the case with the generators. In the end, solar energy does not necessitate that much water for the maintenance purpose. In simple words, a person will end up saving on the water which can be used in order to irrigate the crops.

3.10.2. Solar Energy Is Free

Not like the electricity where a person is obligated in order to make monthly payments, solar energy is free. A person is required to have solar panel which is installed on the form, and then a person is good to go. In addition to this, make sure the person is investing in the best quality of the solar panels that are installed with the help of a seasoned expert.

3.10.3. Solar Energy Is the Perfect Alternative to Electricity

Farmers who use the electricity in order to run their irrigation system know exactly how disrupting an impromptu blackout can be. As the generators are very costly to both run as well as maintain, solar energy is every so often seen as the perfect, which is the most efficient substitute. And the reason behind this is that, it not only runs on the sunlight, which is free, but the system is also quite straight forward from an operational standpoint.

3.10.4. Solar Energy Improves Grid Security

The grid will always be safe guarded as long as it has solar energy penetration.

This explains why a network with power plants that are spread out will have experience very few blackouts. The latter will also play a very important role in safeguarding the grid in the action of a natural disaster, some of which can take place out of the blue.

3.11. DISADVANTAGES OF SOLAR POWER IRRIGATION

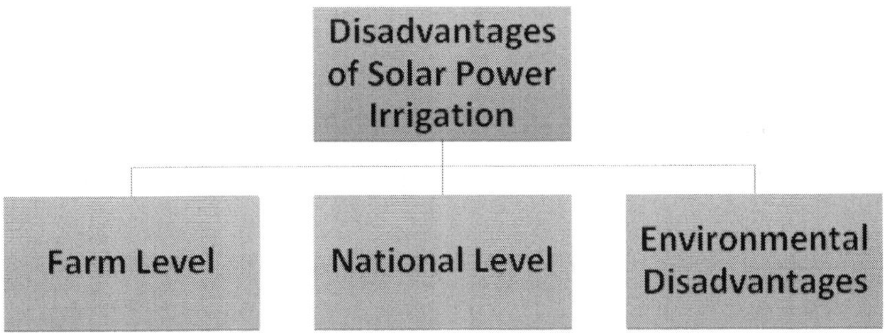

Figure 3.10. Disadvantages of the solar power irrigation.

3.11.1. Farm Level

1. Still comparatively high initial investment costs that smallholder farmers, more particularly, cannot afford or cannot tolerate the dangers come along with the investment; (Figure 3.10).
2. Finance is not accessible or affordable for all, more particularly for the smallholder and tenant farmers;
3. Design require to be fit-for-purpose and necessitates services (generally a private sector) in order to advise farmers on best system; never the less, these are every so often not in place;
4. Optimal operation and maintenance of Solar Power Irrigation System necessitates a certain amount of technical knowledge and skill, in this manner, the farmers are required to be trained and services (extension services or private service suppliers) need to be available.;
5. Solar Power Irrigation System are vulnerable to theft and in this way, they, every so often not covered with the help of insurance as a prerequisite for loan finance; and

6. Deficiency of trust between the farmers, utilities, service providers and government to try innovative forms of finance and of FITs; banks every so often perceive that Solar-Powered Irrigation System have high risk, and the reason behind this is to unfamiliarity with the technology.

3.11.2. National Level

1. Existing energy subsidies for fossil fuels and electricity that distort market; legislation and regulation of energy and agricultural markets may hamper the uptake and up-scaling of solar energy systems. Taxes on imported equipment that may distort (and artificially keep up) prices.
2. High risk investment, more particularly, if the Solar-Powered Irrigation System roll-out programs do not sufficiently address onsite ownership
3. Deficiency of groundwater management/institutional framework for abstraction.
4. Deficiency of codes and standards in order to guarantee the quality of Solar-Powered Irrigation System.
5. Deficiency of systematic training schemes (for example, vocational training); in this way, Deficiency of skilled personnel.

3.11.3. Environmental Disadvantages

1. Production of Photovoltaic panels liberate some toxins as well as rare minerals; mining and production of these tends to cause the environment by producing the environmentally harmful waste; panels are required to be correctly disposed of in order to avoid environmental dangers.
2. Decentralized systems are very challenging task in order to control (never the less, Solar-Powered Irrigation System, every so often substitute already decentralized systems); illegal pumping may elevate.
3. Risk of groundwater over abstraction, that takes into the direction towards the depletion and degradation of groundwater resources.

3.12. CONCLUSION

In the conclusion of the chapter, the agricultural farm is the backbone of almost each and every local economic development, more particularly in the developing nations. And there are three factors in the field of agriculture that are considered as most important, that are soil, water, and sun.

Sun is replaceable by artificial light. Soil is replaceable by the hydroponics. But there is no alternative to the water. Water irrigation is very necessary in the case of deficiency of rain.

There are some important factors that are need to be consider in order to accomplish or achieve farm, for example, reliability, affordability, and access to irrigation water.

More particularly, the farmers who live in rural regions have no access to the reliable source of water for the process of irrigation. And if there is access, they are not able to afford it. One promising solution to the related issue, considering these aspects, is the solar-powered irrigation system.

Solar power irrigation system (SPIS) is an automatic irrigation system. In Solar-powered irrigation system, the irrigation pump is driven with the help of the electric energy that is produced with the help of the solar radiation or sunlight, which is transformed with the help of solar panels or pv cells.

Installation of the solar-powered irrigation system is an expensive investment, but on the other hand, it is actually a cost effective-talking about long term as compared to the conventional water pumps that use non-renewable fossils such as diesel and gas that contributes to further harms of the environment.

This technological option become very familiar and well-known, and the reason behind this is that, a noticeable reduction in the panel prices and the cooperation or help of governments as well as technical extension services.

Solar-powered irrigation system has a certain importance, not only with respect to the economy of a country but most particularly to the environment.

REFERENCES

1. Adnalandal, (2019). Major Benefits of Solar Water Irrigation. [online] Urja Global Limited. Available at: http://urjaglobal.in/major-benefits-of-solar-water-irrigation.php (accessed on 20 August 2020).
2. Electronics for You, (2014). Solar-Powered Irrigation Systems | Electronics for You. [online] Electronics for You. Available at: https://www.electronicsforu.com/technology-trends/tech-focus/solar-powered-irrigation-systems (accessed on 20 August 2020).
3. Energy pedia, (2018). SPIS Toolbox-Solar-Powered Irrigation Systems-Energypedia.Info. [online] Energypedia.info. Available at: https://energypedia.info/wiki/SPIS_Toolbox_-_Solar_-_powered_Irrigation_Systems (accessed on 20 August 2020).
4. Frometa, R. J., (2019). 7 Benefits of Having Solar Power Irrigation-Vents Magazine. [online] Vents Magazine. Available at: https://ventsmagazine.com/2019/09/12/7-benefits-of-having-solar-power-irrigation/ (accessed on 20 August 2020).
5. Green match, (2020). Advantages & Disadvantages of Solar Energy (2020) | Greenmatch. [online] Greenmatch.co.uk. Available at: https://www.greenmatch.co.uk/blog/2014/08/5-advantages-and-5-disadvantages-of-solar-energy (accessed on 20 August 2020).
6. Hans, H., (2018). The Benefits and Risks of Solar-Powered Irrigation—A Global Overview. [ebook] the Food and Agriculture Organization of the United Nations and Deutsche Gesellschaft fürInternationaleZusammenarbeit. Available at: http://www.fao.org/3/i9047en/I9047EN.pdf (accessed on 20 August 2020).
7. Helen, D., (2019). What is Solar Irrigation? Futurepump? [online] Futurepump. Available at: https://futurepump.com/what-is-solar-irrigation/ (accessed on 20 August 2020).
8. Nevon Projects, (2019). Irrigation System Running on Solar Power. [online] Nevon Projects. Available at: https://nevonprojects.com/irrigation-system-based-on-solar/ (accessed on 20 August 2020).
9. Renewable Resources Coalition, (2016). Solar Energy Disadvantages: The Top Drawbacks of Solar Power. [online] Renewable Resources Coalition. Available at: https://www.renewableresourcescoalition.org/solar-energy-disadvantages/ (accessed on 20 August 2020).
10. Solar Power Nerd, (2020). What is A Solar-Powered Irrigation System? | Solar Power Nerd. [online] Solar Power Nerd. Available at: https://

solarpowernerd.com/what-is-solar-power-irrigation-system/ (accessed on 20 August 2020).

11. Stéphanie, R., (2016). Solar-Powered Irrigation: A Solution to Water Management in Agriculture?-Renewable Energy Focus. [online] Renewableenergyfocus.com. Available at: http://www.renewableenergyfocus.com/view/44586/solar-powered-irrigation-a-solution-to-water-management-in-agriculture/ (accessed on 20 August 2020).
12. Tarun, A., (2020). Solar-Powered Automatic Irrigation System-ECE Projects. [online] ElProCus - Electronic Projects for Engineering Students. Available at: https://www.elprocus.com/solar-powered-automatic-irrigation-system/ (accessed on 20 August 2020).

Chapter 4

Micro-irrigation Management

CONTENTS

The micro-irrigation system has developed as one of the most effective irrigation systems in the modern world. This chapter discusses the importance of managing the micro-irrigation system. This chapter also discusses about the concept of nutrigation, which focuses on distributing the nutrients effectively to the plants during the process of irrigation. After discussing all the major, secondary, and micro-nutrients, this chapter focuses on drip irrigation as the sub-division of micro-irrigation system.

Drip irrigation is the most extensively used micro-irrigation technology. Along with the design criteria, characteristics of fertilizers and the attributes of drip irrigation system are discussed in this chapter. Also, the subsurface drip irrigation is discussed in this chapter along with its design criteria, filtration requirements and valves used in the subsurface drip irrigation system.

4.1. INTRODUCTION

Drip irrigation systems and sprinkler irrigation systems have experienced an immense increase in commercial significance. It has also been observed that there was an increase in the research interest in the last three decades.

Both these systems are the classification of the micro-irrigation system. These systems are well-known in the modern world. One of the major reasons for these systems becoming so famous is that they have become more realistic and practical with the development of plastics. The eco-friendly plastics are inexpensive, durable, and easily molded into the variety and complexity of shapes required for pipes and emitters.

The systems that were used in the earlier times were made up of small-bore (1–2 mm ID) plastic microtubes (emitters). This was inserted inside the plastic pipe which was of larger-diameter. The rate of water emission was controlled by friction in the microtube. It produces the irrigation rates of 2 liters to 6 liters each hour per emitter.It has been observed that the operating pressures were low. These pressures were often about 103 kPa. This requires a careful system design in order to offset the impacts of small elevational changes on the emission rate and the internal system pressure.

Later, the emitters have advanced into various forms like from discrete emitter units to several kinds of porous hose, and in the recent times, it has evolved to pressure-compensating types. This type of emitters regulate flow over a range for pressures, allowing the higher pressures to be used in drip irrigation lines.

There have been several recent innovations that were designed in order make the micro-irrigation adaptable to different conditions. This consists of moving units, simplified bubbler systems, and the promotion of spitters or micro-sprinklers in place of drippers.

It is possible that one of the most significant trends in micro-irrigation management has been in the direction of attaining a greater control and controlling the automation of the frequency in the irrigation. Also, the amount of water used for irrigation is managed with the use of programmable computer-based systems.

Along with computer-based programs, management of micro-irrigations systems also focuses on a number of devices such as sequential metering valves and sensors to monitor weather and soil moisture variables.

In the recent times, innovative or new drip-irrigation systems have been introduced for usage with wastewater in both garden setting and agricultural settings. There is a labyrinth "toothed" water passage in the rubber tubes of these pipes. This helps in facilitating better filtration. It has been observed that chemigation and especially fertigation are yet other developments of main significance.

There has been a rapid expansion in the plantation area that is irrigated by the drip irrigation method. The first international survey was conducted by Gustafson et al. (1974). A team of scientists led by Gustafson indicated that near about 56,000 ha were under micro-irrigation all over the world.

Also, it was observed through this study that almost half of the area irrigated by micro-irrigation systems was present in the USA. According to a recent survey (Abbott, 1988), it was indicated that there was an increase in the area, that was irrigated by drip irrigation alone, was expanded up to 1,081,000 ha.

There have been several references all over the world where micro-irrigation systems have been used as the only irrigation system. One of these countries is Cyprus. In Cyprus, drip, and mini sprinklers are used to irrigate more than 95% of the irrigated land. The majority of the irrigated land is occupied by orchards and vegetable crops, unlike the conventional field crops occupying the majority of the irrigated land.

This is major because of the higher unit-crop value of the orchard and vegetable crops. Also, the field crops require an excessive irrigation system which increases the cost factors associated with irrigation for crops.

4.2. MANAGING THE DESIGNS AND LAYOUTS OF MICRO-IRRIGATION SYSTEMS

It has to be noted that the uniformity of water usage is essential in order to maximize the growth of the crop growth and the efficiency of water use. The differences in pressure over the system impacts the uniformity of the water application. Here, the designs and layouts of drip irrigation system are being discussed (Figure 4.1).

Figure 4.1. Managing micro-irrigation systems helps in understanding the entire process.

Source: Image by Flickr.

Usually, the drip systems function at pressures between 8 psi and 12 psi, and the differences in pressure as low as 2 psi can significantly decrease the uniformity of application. There are many other factors that have impact on the uniformity of water application. These include, plugged emitters, tape splicing, the use of tapes that are different flow rates, and sloped fields.

The usage of drip tape with pressure-compensating emitters and the division of the fields into zones of irrigation can helps in minimizing some of these issues.

4.2.1. Drip Line Arrangement

The arrangement of drip lines within the planting is one of the variables that is required to be considered when laying out a drip system. It has been observed that there can be a single drip line for every two rows (every other row), one line for each row, or two lines for each row.

From the plant row, the distance of the drip line can also differ, with spacing's of 6, 12, and 24 inches. These spaces are being the most common. The number of lines that are used are dependent on the properties of soil, economics, crop water demands and grower expertise and preferences.

In addition to it, spacing choices also includes compromises between the cost of the system and optimal water distribution. It has been observed that at least one line for each row helps in providing optimal water distribution, particularly in sandy soils. On the other hand, one line every other row can help in providing sufficient delivery of water at a lower price. It is provided on heavier soils with good horizontal water dispersal.

It has been observed that water moves down through the soil profile much quicker than it moves horizontally in case of sandy soils. There is the need of a closer spacing of drip lines. On the other hand, in the case of clay soils, water might move laterally faster than it moves in the downward direction. There is a need of wider spacing of lines. It might help in deeper root growth.

It is beneficial to have two lines per row for the widely-spaced, high-demand crops, like watermelon. In addition to it, it is also beneficial for all those crops that are grown on sandy soils. Higher soil moisture levels in the top segment of the root zone is generally caused by putting drip lines close to the crop row. But it has to be noted that it does not always result in product quality and higher yield. This is because putting the drip lines further from the crop row can help in deeper root growth.

4.2.2. Buried vs. Surface Line Placement

Drip lines can be positioned on the soil surface (that is planting bed) or it can also be buried under the surface of the soil. It has been observed that the placement of drip line on the surface of the soil that is the Surfaced placement can be easier and it also helps in the maintenance, easy inspection, repositioning, and removal of the drip lines from the surface at the end of the season.

Although, drip lines that are placed on the surface of the soil tend to move with the speed of the wind and these lines are more susceptible to damage by the equipment, animals, and foot-traffic. On the other hand, the drip lines can be buried 1 to 3 inches under the surface between plant rows or it can be placed under shallow-rooted crops (that is onions).

Placement of drip lines at depths of 6 inches to 12 inches can be used for crops that are deep-rooted. On the other hand, placement of drip lines at the depths of up to 18 inches is used in the case of permanent subsurface drip irrigation (SDI) systems.Drip lines that are buried under the soil surface will not move, and they are better guarded from damage that can be caused because of any reason such as animals, equipment. Though, the maintenance as well as the removal of the drip lines are more difficult and also, costly.

Also, the subsurface drip line systems do not wet the soil surface, and this acts as an advantage. As the soil surface is not wetted, it eliminates the loss of water through evaporation. Along with preventing the water loss, subsurface drip line system makes the water dispersal more uniform within the soil.

4.2.3. Mulched vs. Bare-Ground Irrigation

It has been observed that the rate of evaporation from plastic-mulched soil can be half as compared to that of non-mulched soil. Though, the rate of growth of plants in mulched-bed systems can be faster as compared to those plants that are growing in bare soil, and the larger plants will make use of more quantity of water per unit area. Even so, the total quantity of water that is utilized per unit of production will be less in the case of a mulched system.

The growth of weed is also suppressed by opaque plastic mulches and it also helps in reducing the usage of water by weed plants. This further helps in enhancing the efficiency of water use. Though, plastic mulches help in preventing the rainwater from reaching the root zone, and in most of the cases, rain-water does not contribute significantly to the water requirements of the crop in plasticulture systems.

4.3. NUTRIGATION-MANAGING THE NUTRIENTS

There is a need of balanced diet of important nutrients throughout the growth cycle for the crops. There are various plant foods that can be found in the soil, but these are often found in inadequate quantities to sustain the high crop yields (Figure 4.2).

Figure 4.2. Nutrigation is the process of providing nutrients along with water that is sent through irrigation system.

Source: Image by Wikimedia Commons.

Soil as well as the climatic conditions can also be the reason of limiting the plant's uptake of nutrients at the important stages of growth. There are various crop scientists who recognize that plants require 13 important minerals. All these important minerals play a number of essential functions. If any of these minerals is lacking in the plant then the growth of the plant and yield suffer.

4.4. MANAGEMENT OF MAJOR NUTRIENTS

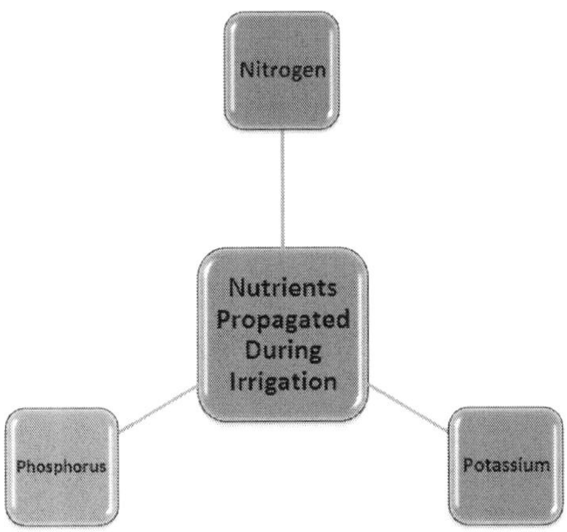

Figure 4.3. Major Nutrients propagated during irrigation

4.4.1. Nitrogen (N)

Nitrogen is often needed by the crops in the greatest quantity, mainly for the purpose of yield and vigor. It has been observed that Nitrogen plays a very important role in the production of chlorophyll and protein synthesis. Chlorophyll is described as the green plant pigment that is responsible for the process of photosynthesis. The plants develop yellow leaves or pale leaves and their growth is also stunted when nitrogen is deficient (Figure 4.3).

4.4.2. Phosphorus (P)

Phosphorus an important component of adenosine triphosphate (ATP). This component supplies the energy to the plants for various processes. It has been observed that phosphorus hardly produces spectacular growth responses but is important for the effective development of all the crops. For instance, when there is lack of phosphorus in maize or other corn crops during the growing season then these crops achieve lower yields.

4.4.3. Potassium (K)

All crops need potassium and often they need it in higher rates as compared to that of nitrogen. Potassium helps in regulating the water content as well as expansion of the plant. It is very important to achieve good quality as well as yield in cotton and essential for increasing the content of juice, size, and sweetness of the fruit.

4.5. SECONDARY NUTRIENTS IN MICRO-IRRIGATION MANAGEMENT

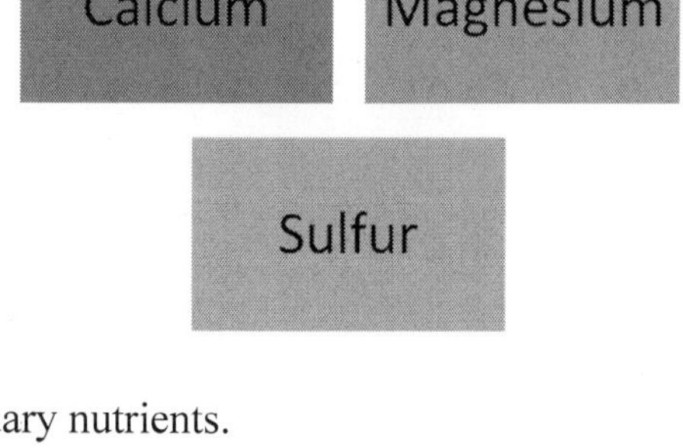

Figure 4.4. Secondary nutrients.

4.5.1. Calcium (Ca)

Calcium is the secondary nutrient, but it is one of the most important nutrients. Calcium helps in strengthening the walls of the cell, helping to decrease the bruising as well as disease in salad, fruit, and vegetable crops. It means that the supply of calcium in the good quality produces food crops that are less susceptible to damage and they have a longer shelf life. On the other hand, those crops that lacks calcium will have growth disorders like corky skin (Figure 4.4).

4.5.2. Magnesium (Mg)

Magnesium is vital for the quality of the crop, and it is also a very important element of leaf chlorophyll and the enzymes that helps in supporting the growth of the plant. It has been observed that low magnesium or the lack of magnesium is the reason of reduced photosynthesis. It limits the crop yields. For instance, dry matter content of potatoes and grain fill in rice can be drastically reduced if there is lack of magnesium in it.

4.5.3. Sulfur (S)

Sulfur is an important part of many proteins and amino acids. It has been observed that in the absence of both sulfur (S) and magnesium (Mg), crops suffer; the growth of the crops slow down, and the leaves of the crop turn yellow or pale. In addition to it, sulfur is very much important especially for ensuring the content of protein of cereal crop grains.

4.6. MANAGING THE MICRONUTRIENTS

Micronutrients are the nutrients that are needed by the plants to reinforce and supplement the major and secondary nutrients. The micronutrients enhance and support the growth that is imparted by the major and secondary nutrients (Figure 4.5).

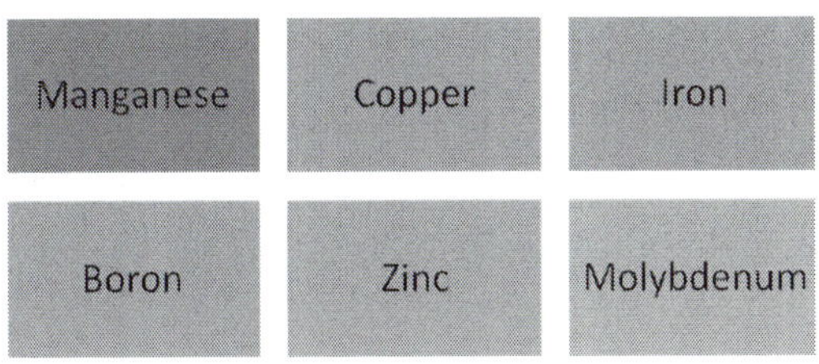

Figure 4.5. Some of the micro-nutrients propagated during micro-irrigation.

4.6.1. Manganese (Mn)

Influences the process of photosynthesis in the plants. It is defined as the process by which plants make use of sunlight for the purpose of their growth.

4.6.2. Copper (Cu)

It also helps in influencing the process of photosynthesis. This process is very much important for the growth of the plants.

4.6.3. Iron (Fe)

Iron also Influences the entire procedure of photosynthesis. Deficiencies are common, for instance, in seed fruits-where the effect is to increase the production of chlorophyll. As a result of this, crops suffer, and younger leaves turn in pale color or severe yellowing or it can also be called as chlorosis.

4.6.4. Boron (B)

Boron is required for the development of roots and shoots. It is important during the fruiting and flowering stages of the crops.

4.6.5. Zinc (Zn)

Zinc (ZN) is required in order to produce the important plant hormones, such as auxin. It has to be noted that the deficiency of Zinc leads to structural defects or deficiencies in leaves and other parts of the plant. It is very much important for the proper growth of plants.

4.6.6. Molybdenum (Mo)

Molybdenum (Mo) is involved in plant enzyme systems. It helps in controlling the nitrogen metabolism.

4.7. NUTRIENTS AS MAJOR REQUIREMENTS FOR CROPS

There is the requirement of different variety of nutrients for every single plant at every important phase of its development. For instance, the nutrients such as phosphorous and nitrogen are often more important at initial phases of growth. These are required at the early phases to fuel root as well as leaf development. On the other hand, the nutrients such as zinc and boron are essential for the plants during the phase of flowering.Cereal crops make use of nutrients for the purpose of their growth, gradually moving them from the roots of the plants, to the stems and leaves until the remains of the plants dying off and harvesting of the grain. Tree crops have different nutrient requirements than field crops.

The nutrients such as nitrogen can be stored within the trunk, leaves, and branches of the plants. Then this stored nitrogen can be redistributed to the important points during the cycle of growth. However, it is essential to provide the trees with replacement levels of the nutrients that are removed while harvesting the fruit.These nutrients are critical and important for the growth of the tree. The nutrients that are depleted from the trees during the fruit harvesting cannot be recycled. Therefore, there is always a need to replenish those nutrients into trees.

4.8. USING DRIP IRRIGATION SYSTEM FOR NUTRIGATION

Nutrigation is defined as the combined application of nutrients as well as water to a crop. It is a mixture of irrigation and fertilization. It has to be noted that it can be adapted to all the kinds of crops.

While using the drip irrigation system, the roots are concentrated in a limited soil volume as compared to the volume of soil that these roots occupy if the irrigation is not localized (Figure 4.6).

Figure 4.6. Drip irrigation eradicates the water losses while watering the roots of the plants.

Source: Image by Wikimedia Commons.

Because of this reason, during the season of irrigation, comparatively small quantities of fertilizers should be frequently used. In traditional fertilization methods, some quantity of the fertilizer is applied outside the volume of soil which is occupied by the roots. Eventually, this leads to wastage of fertilizers.

4.9. ADVANTAGES OF USING A DRIP IRRIGATION SYSTEM

There are various advantages of managing the micro-irrigation. The drip systems, used in micro-irrigation, can help in reducing the use of water by 50% as compared to that of the sprinkler irrigation. Also, the requirements of lower pressure can result in reduced pumping costs.

It has been observed that often, yields are higher in cropping systems when drip irrigation is used. This is majorly due to improved water management and diminished spread of disease and negligible weed growth. It has been observed that the wetting of foliage takes place with sprinkler irrigation. And, this can result in the spread of some plant pathogens and it can also cause some diseases. But this is not the case with drip irrigation, and drier plant canopies can help in improving the aspects relating to the concerns of food safety.

In Drip systems, there are high water use efficiency (WUE) levels. This is because water is directly applied to the root zone and it is not provided to nonproductive areas like the spaces between rows.

It has to be noted that the drip systems are very adaptable. They can be even used in those fields that are having uneven topography and also, in unusually shaped fields. In addition to it, drip systems can also be used to provide pesticides (chemigation) as well as nutrients (fertigation) to the crop during the season without interrupting the damaging foliage or the root systems.

It has been observed that in some of the cases, plastic mulch and drip-tape are left in place and these are used again for one or more following crops (for example, cucumber following tomato).

4.9.1. Increased Efficiency of Fertilizer Application

- While using a drip irrigation system, the total amount of fertilizers is divided into several smaller portions. This allows the soil to retain the fertilizers in a better way and increase the availability of fertilizers.

- Application of fertilizer while irrigating the plantation area helps in providing a better distribution of fertilizer while ensuring greater application uniformity.
- Losses because of the volatilization are prevented, as with fertilizers that contain the compounds of nitrogen.
- Nutrigation allows the application of nutrients as per the crop requirements. Also, there is a possibility that there might be a variance in the ratio between requirement of nutrients in various phenological stages, such as the vegetative stage, flowering, fruit setting, fruit filling and maturation.

4.9.2. Savings in Fertilizer and Labor

- The amount of fertilizers being used in the plantation area is considerably reduced due to the conditions of application by irrigation water. With proper irrigation systems in use, higher yield is achieved with same level of fertilizers.
- The production and application of fertilizers cost much less in micro-irrigation system as compared to the traditional irrigation methods.

4.9.3. Operational Advantages

- Health hazards are prevented since the people who are working do not come in contact with the injected chemical and the fertilizers;
- Since there is no need of traffic for Nutrigation in the field, soil compaction and damage to plants and are avoided;
- Nutrigation can maintain proper content of nutrient in soils with low nutrient holding capacity. This allows the farmers and plantation owners to cultivate in those types of soil in which cultivation would not have been possible;
- The contamination of groundwater by elements of fertilizer occurs in many places where flood irrigation is used. Nutrigation applies the amount of fertilizer and water in more frequent, smaller portions to prevent runoff or deep percolation; and
- There is the possibility of applying other chemicals through the system, such as soil disinfectants, systemic products against crop diseases and pests.

4.10. LIMITATIONS OF USING DRIP IRRIGATION SYSTEM

- In drip irrigation, only the water-soluble fertilizers can be used;
- There are some of the fertilizers, even though water soluble, may not be compatible with the method of Nutrigation™, like those fertilizer that increase the pH of the irrigation water so high that precipitation take place in the system; and
- There are some fertilizers that are corrosive to metal parts of the equipment. Thus, the parts of the system that is coming in contact with these kinds of fertilizers should be resilient to corrosion.

There are various shortcomings or disadvantages of drip systems. These disadvantages consist of higher degree of skill and management, a high initial investment (that is around $1,200 to $2,000 per acre). There is the need to monitor Drip systems on the regular basis for clogs, damage, and leaks so that repairs can be quickly made without the wastage of any time.With the temporary drip installations (that is common with most of the vegetable crops) the drip tape is required to be removed at the end of the season. This results in costs of removal of drip tape and problems with disposal of the tape that is used.There is a need of a higher level of water quality for Drip systems as compared to any other irrigation systems. There is the need of clean water within a specific range of pH and mineral content in order to reduce the problems with clogged emitters.In addition to it, there is the need of multiple filtration systems to adequately clean the water, and also, periodic acid and chlorine flushes may be required in order to remove the problems with mineral (iron) deposits and the growth of algae as well as bacteria that can clog emitters.Because the distribution of water is limited to the root zone, salt accumulation can become an issue in some of the areas. This depends on the occurrence of natural rainfall and on the quality of the water.

4.11. CHARACTERISTICS OF FERTILIZERS USED IN NUTRIGATION

4.11.1. Chemical Composition

It has to be noted that fertilizers can be compound or simple or compound:

- Simple fertilizers: Simple fertilizers are those kinds of fertilizers that includes a single product. For instance: potassium chloride, urea, ammonium nitrate;

- Compound fertilizers: Compound fertilizers are the products that are acquired by mixing various simple fertilizers, and usually, it can be seen easily in the mixture. These kind of compound fertilizers are not generally used in Nutrigation;
- Solid state: The fertilizer may be powdered or granulated in the solid state; and
- Liquid state: In this case, there are fertilizers that can be directly injected into the irrigation system. There are some fertilizers that are required to be dissolved in water in order to decrease the concentration before the fertilizer is injected.

4.11.2. Solubility

Solubility is considered as one of the essential characteristics that is to be considered in the preparation of liquid fertilizers. It has been observed that every single fertilizer has a level of solubility. This level of solubility is influenced by the temperature of the water in which it dissolves.

4.11.3. Interaction with Irrigation Water

Fertilizers are defined as the salts that react with other salts that are found in the irrigation water. Thus, it is essential to consider the chemical composition of the water that is used for the preparation of liquid fertilizers. For instance, the phosphorus of a phosphate fertilizer precipitates with magnesium and calcium that is present in the water, under conditions of high alkalinity water. These precipitates are seen at the bottom side of the fertilizer tank.

4.11.4. Interaction Between Fertilizers

It has been observed that there are some fertilizers that are not be mixed on the same mixture because they are incompatible. In some of the cases, when these fertilizers are mixed, those fertilizers instantly induce crystallization. In addition to it, it also causes clogging in the irrigation system. On the other hand, in some other cases the reaction that take place between two fertilizers that are incompatible in nature, causes the loss of nutritious ingredients to the plant.

4.11.5. Corrosivity

It has been observed that most of the fertilizers, both liquid and solid, attack metals in the fertilization system and irrigation systems. It has to be noted that the higher the acidity of the solution, the greater is the corrosive effect. For instance: the mixture of phosphoric acid and potassium chloride is extremely corrosive in nature.

4.11.6. Volatilization

Fertilizers that are having ammonium nitrogen and urea can be lost by the process of volatilization of ammonia. The fertilizer tanks that are keeping liquid fertilizer mixtures for a longer period of more than 4 days must be sealed or closed.

4.11.7. Salinity

Fertilizers are basically the salts that helps in contributing to the increased salinity of the water. Salinity is usually the amount of salt that is dissolved in water. The salinity of the water is reflected by the level of EC that is electrical conductivity. And this salinity in the irrigation water is calculated with the help of simple instruments that are available in the laboratory and in the field.

4.11.8. Hygroscopicity

It has been observed that the solid fertilizers have the property of adhering to moisture. This property of the fertilizers stiffens or solidifies the granules. This, in turn, makes the handling of granules more difficult. It is essential to keep them in a container that is closed or sealed. This is done in order to prevent this phenomenon.

4.12. SUBSURFACE DRIP IRRIGATION

Subsurface drip irrigation (SDI) method is proving to be a cost-effective method of water application to agronomic row crops like cotton, peanuts, and corn. It has been observed that many advantages are provided by the SDI as compared to the other systems of irrigation (Figure 4.7).

Figure 4.7. Subsurface irrigation system is used to ensure effective use of fertilizers as well as chemicals.

Source: Image by Public Domain Files.

It has been observed that in this system of irrigation, there is an increase in life expectancy and less amount of annual labor involved. A dry surface of soil helps in reducing the occurrences of soil-borne infections and it also helps in controlling weed infestations. In addition to it, the dry soil in furrow helps in improving the traffic ability and it also decreases the soil compaction. There is a more effective utilization of water as well as nutrients, and in addition to it, there is a significant enhancement in yield as well as quality components.

Usually, SDI is defined as a "permanent" system. That means the drip lines are not taken up every single year. SDI systems must be designed and installed very carefully so that they function with proper effectiveness and so that the fertilizers as well as the chemicals can be used in a uniform and effective manner.Before the completion of the structure or design of an SDI system, it must be determined that the planned site is appropriate for SDI. These considerations consist of adequate supply of water, acceptable quality of water, and proper topography. The other important consideration is management. This consideration is vital for all drip irrigation systems and particularly important to the SDI systems in which the drip lines are under the ground that is these drip lines are out of sight.

It has been observed that the design of an SDI system is same as the design of other drip irrigation systems but in SDI system, there is some additional consideration given to traffic and system flushing. Topography and field layout will usually demand more attention in humid regions.

4.13. DESIGN CRITERIA

4.13.1. Water Requirements

The design of the SDI system must be proper so that it deliver the necessary flow rate, and the supply of water must be sufficient to deliver the amount of water that is needed over the growing period. The SDI system may be planned to irrigate more than a single crop (rotation).

In this case, the crop that is having the highest demand of water should be satisfied. The quantity of water that is needed will be dependent on various factors, which includes crop, climate, and soils.

Climate: It has been observed that in humid areas, the requirement of water is usually lower than more arid regions. This is because of the reduced driving force or vapor pressure gradient, for evapotranspiration. There are various other factors like sunshine, temperature, and wind that also have impact on evapotranspiration. With the help of an SDI system, it has been observed that evaporation from irrigation is reduced to a small amount in majority of the cases, since the surface of the soil is not usually wetted.

Crop: The vital part of crop water use for SDI design is the "peak" requirement of water or the amount of water that is used by the crop at the time of its highest water use period.

Soils: The type of the soil into which an SDI system will be fitted can also have influence on the system design. The characteristics of soil such as structure, texture, and layering can also have impact on soil hydraulic characteristics like hydraulic conductivity and infiltration rate.

The drip lines are required to be more closely spaced in case of sandy soil since the lateral spread of water from the drip lines will be less pronounced as compared to that in a finer textured soil. There is the need of slow emitter emission rates on heavy textured soils, like clay, so that the rate of emission does not surpass the hydraulic conductivity of the soil.

Generally, the depths of drip line should be shallower in case of coarser textured soils. On the other hand, the depths of drip lines should be deeper in case of finer textured soils.

4.13.2. Management and Operation Considerations

The consideration of how the use of the area will differ, is the most important step in the design of an SDI system. This system must not focus only for the crop yield for the next year (Figure 4.8).

Figure 4.8. Micro-irrigation systems are sustainable and are installed so that they can function for at least 15 years.

Source: Image by Flickr.

A system that is designed in a proper way or well-designed should be in function for at least 10 years to 15 years, so there is the need of some attempt that should be made in order to plan it for the future. There are few important questions that should be asked during this phase. These includes:

a. Will the same type of crop be grown every single year, or will there be a rotation of various crops?

b. Is sub-soiling a part of the production system?

c. Will the whole field be planted to a single crop or will it be separated into smaller areas of various crops?

d. Will the different crops in a rotation make use of different cropping systems?

Along with these considerations, there are a number of system maintenance operations such as, iron settling, acid injection, flushing, etc. These operations are not necessary or are less extensive than for some of the other types of irrigation systems. Regular maintenance must not be avoided. Avoiding or ignoring the need of maintenance of the systems might result in significant decrease in the system life and thus, will lead to collapse of the entire system.

4.13.3. Water Quality

At the time of designing any type of irrigation system, water quality concerns may involve two sources of design criteria, one is for the crop and the other is for the system. Usually, it has been observed that the water quality criteria for crops emphasizes on leaching requirements or application concerns (foliar burning, etc.).

A leaching requirement is not required in humid areas, where salts do not build up in the root zone. Also, since it has been observed that SDI systems don't wet the plant so the issues that are resulting from the contact of irrigation water with the plant are not a problem. As a result of this, the water quality criteria for the design of SDI systems in humid regions pays attention on the concerns about the irrigation system.

It has been observed that emitter clogging is the main concern with SDI systems along with every kind of drip system. The quality of the water will dictate chemical injection requirements, filtration requirements and management of SDI systems to avoid the problem of emitter clogging.

4.14. FILTRATION REQUIREMENTS FOR SUBSURFACE DRIP IRRIGATION

In SDI system, filtration is very important. The location of clogged emitters is very hard to find since the emitters are buried. It should be kept in mind that only clean (that is free of particulate matter) water should be pumped through drip irrigation systems for optimum performance.

The size as well as the kind of filter that is needed will be dependent on the source of water and the types (if it is present) of fertilizer and chemical stock solutions to be injected. It has been observed that these filtration systems are positioned at the headwork's of the SDI system.

1. Media Filters: Media filters are considered to be superior for filtering the surface water because of their large capacity and filter area. In fact, the media filters can be easily back-flushed for cleaning. Moreover, this entire system can be automated. During the back-flushing operation, there is the need of at least two media filters, clean water from one of the filters is used in order to remove pollutes from the other filter.
2. Screen Filters: These are used as secondary filters with surface water systems, or these are used as primary filters with well or municipal sources of water. Screen filters are different in both the size and the shape. It has been observed that these screen filters may differ in size from 1 inch (that is for one acre or less) to 10 inches in diameter. Usually, the mesh sizes range from 40 to 200. Even though there are some screen filters that are having automatic back-flush capabilities, most of them are required to be flushed manually.

3. Disk Filters: Disk filters are used as primary filters with well or municipal sources of water, or they are used as secondary filters with surface water systems. These filters consist of a series of grooved plastic disks that might have an equal screen size. The size varies between from 40 mesh to 200 mesh. It has been observed that disk filters have the benefit of having more surface area as compared to that of screen filters. Thus, disk filters are well suited for higher flow rates and in addition to it, these are also easy to clean.

4.14.1. Chemical Injection for Subsurface Drip Irrigation

Chemigation indicates the use of a chemical in the irrigation system. It consists of the use of acids, fertilizers, chlorine, and pesticides. It has been observed that drip emitters clog easily. This is because they are small. Along with the process of filtration, the ability to inject acid and chlorine are the essential features in an SDI system.

This is found particularly in humid areas for the purpose of removing algae. There are several other advantages of chemigation. These benefits are uniform as well as timely usage of fertilizer, reduced requirements of labor, reduced soil compaction because of the reduced traffic, exposure to chemicals, and environmental pollution.

There are various factors that are involved in the design of a chemical injection system. These includes the choice of the injector, both capacity (that is size) and type.

In case, if the injection system is supposed to be used for fertigation, then the injection unit must be designed according to the size for that particular area. This is because the injection rates for fertilizers are generally much higher than injection rates for chemicals such as liquid chlorine or acid. There are two basic kinds of injection pumps. These are the metering pump and the Venturi injector. These are used commonly for the injection of fertilizer as well as some other chemicals into the drip irrigation systems.

4.15. VALVES FOR SDI SYSTEMS

The proper placement and selection of valves are very important in an SDI system. The rate of water flow and pressure throughout the SDI system should be controlled precisely in order to ensure timely as well as effective water application.

Valves play a very important role in controlling the flow, pressure as well as distribution under different conditions. This is done in order to facilitate management, improve the performance, and reduce the requirements of maintenance in SDI systems.

Valves that are used in a complete SDI system consists of shut-off valves, electronic remote-control valves, check valves, pressure relief valves, pressure regulators, and air or vacuum pressure regulators.

4.15.1. Potential and Applications of Micro-Irrigation

Under conventional flood methods of irrigation, the WUE is very low. This is majorly because of the substantial conveyance and loss of water while distribution. Acknowledging the fast decease of irrigation water potential and growing need for water from various sectors, several number or programs and demand management strategies have been introduced in order to save water and improve the current WUE in agriculture.

It has been observed that one such method had been introduced recently many of the developing countries such as India. This method is micro-irrigation. Both drip irrigation and sprinkler method of irrigation are included in it.

Micro-irrigation (MI) is proved to be an effective method that helps in saving water as well as increasing WUE as compared to the other traditional surface method of irrigation. In the conventional surface method of irrigation, the WUE is just about 35% to 40%.

Many studies have been conducted in various countries that have confirmed that irrigation plays a very important role in increasing the use of yield increasing inputs and improving the productivity of the crop and cropping intensity. It has also provided many advantages to the farmers.

Apart from providing advantages to the farmers, irrigation development also helped in increasing the employment opportunities as well as the wage rate of the agricultural landless laborer, both of which are important to decrease the rate of poverty among the landless labor households. However, it has been observed that water is becoming increasingly scarce all over the world because of several reasons.

The issues of water scarcity are expected to be aggravated further with the decline of irrigation water potential and continued growth of population as well as economic activity in most of the nations that are situated in arid regions and semi-arid regions.

Macro-level estimate was carried out by the International Water Management Institute (IWMI), Colombo. It shows that one-third of thc population all over the world would face the issue of absolute water scarcity by the year 2025 (Seckler, et al., 1998; Seckler, et al., 1999).

According to this estimate, the areas that will be affected the most will be the Middle-East and Sub-Saharan Africa and semi-arid regions of Asia. All these regions are already having heavy concentration of population that is living below the poverty line.There are several developing countries such as India that have started facing severe water scarcity in various regions despite of having the largest irrigated area in the world. It has been observed that the demand for water for various purposes has been continuously increasing in many of the developing countries because of several reasons. But the potential water that is available for future use has been decreasing at a very fast rate (Saleth, 1996; CWC, 2005).

In many of the developing countries, the agricultural sector (irrigation) that is presently making use of more than over 80% of the water that is available, continues to be the main sector of water-consuming because of the intensification of agriculture (Iyer, 2003).

The predominant use of flood (conventional) method of irrigation is the primary reason for the low coverage of irrigation. In this, the WUE is very low because of so many reasons.

Many of the existing estimates show that the WUE under flood method of irrigation (FMI) is just about 35% to 40%. This is due to huge conveyance as well as the distribution losses (Rosegrant, 1997; INCID, 1994).

Under micro-irrigation, which is different from FMI, water is provided by the use of emitters, nozzles, and pipe network at the required interval and quantity.Thus, distribution losses and the conveyance and are completely decreased. This result in higher WUE under micro-irrigation. Although sprinkler irrigation methods (SIMs) of irrigation and drip irrigation are treated as MI, there are different characteristics differences between both of them. These differences are in the terms of pressure requirement, wetted area, flow rate and mobility (Kulkarni, 2005).

The water is supplied directly to the root zone of the crop in the drip irrigation method. The water is supplied through a network of pipes with the help of emitters. On the other hand, in SIM, sprinklers are used to spread water in the form which is similar to rainfall. This water is sprayed into the air through nozzles.

The nozzles subsequently break every water drop into small water droplets. Drip irrigation method, unlike the flood irrigation method, provides the water directly to the root zone of the crop, instead of supplying it on the land. Thus, the drip irrigation system helps in minimizing the water losses that are taking place through the process of evaporation and conventional water distribution methods.The on-farm irrigation productivity of the drip irrigation system, that is designed and managed properly, is estimated to be approximately 90%. On the other hand, the same is just about 35% to 40% in case of surface method of irrigation (INCID, 1994).

It has been observed that water saving is comparatively low (up to 70%) in SIM as compared to that of the drip irrigation since the water is supplied by SIM to the entire field of the crop (Kulkarni, 2005).

Micro-irrigation is mainly introduced for the purpose of saving water and increasing the WUE in the agriculture. In addition to it, there are also several other social benefits and economic benefits to the entire society.

The decrease in the consumption of water because of the drip method of irrigation over the surface method of irrigation differs from 30% to 70% for various crops (INCID, 1994; Postal, 2001). In accordance with the information or data that is available from research stations, the increase in the productivity due to the use of micro-irrigation is projected to be in the range of 20% to 90% for various crops (INCID, 1994).

Apart from all this, there are several other benefits. It also helps in reducing soil erosion, the weed problems, and cost of cultivation substantially, particularly in labor-intensive operations. It has been observed that the reduction in the consumption of water in micro-irrigation also decreases the use of the energy (that is electricity) that is needed in order to lift water from the irrigation wells.

4.16. CONCLUSION

Micro-irrigation system is one of the most efficient irrigation systems when it comes to providing the water and nutrient to the plants while sustainable using the resources. It is very important to maintain and manage the micro-irrigation systems. This is so because without proper management and maintenance, the micro-irrigation systems could lose their efficiency and would not last as long they can be. Thus, it becomes very essential to properly manage the micro-irrigation systems and design and use the elements according to the requirements of the systems as well as the plants.

REFERENCES

1. (2015). Drip Irrigation Handbook Understanding the Basics (2[nd]edn.). [ebook] NETAFIM™. Available at: https://www.netafim.com/499749/globalassets/products/drippers-and-dripperlines/drip-irrigation-system-handbook.pdf (accessed on 20 August 2020).
2. Agriculture.vic.gov.au. (2017). Automatic Irrigation | Irrigation | Soil and Water | Farm Management | Agriculture | Agriculture Victoria. [online] Available at: http://agriculture.vic.gov.au/agriculture/farm-management/soil-and-water/irrigation/automatic-irrigation (accessed on 20 August 2020).
3. Cotton Incorporated, (2020). Irrigation Management-Irrigation Systems Overview-Cotton Incorporated. [online] Available at: https://www.cottoninc.com/cotton-production/ag-resources/irrigation-management/irrigation-systems-overview/ (accessed on 20 August 2020).
4. Ecoursesonline.iasri.res.in. (2020). MI&CAD: Lesson 2. Scope and Applications of Micro-Irrigation. [online] Available at: http://ecoursesonline.iasri.res.in/mod/page/view.php?id=199 (accessed on 20 August 2020).
5. Fao.org. (2020). Chapter 6.Drip Irrigation. [online] Available at: http://www.fao.org/3/s8684e07.htm#6.3%20operating%20drip%20systems (accessed on 20 August 2020).
6. Papadopoulos, I., (1996). Micro-irrigation systems and fertigation. Sustainability of Irrigated Agriculture, [online] pp. 309–322. Available at: https://link.springer.com/chapter/10.1007/978-94-015-8700-6_18 (accessed on 20 August 2020).
7. Papadopoulos, I., Metochis, C., &Seraphides, N., (2020). Fertigation Recipes for Selected Crops in the Mediterranean Region. [ebook] API. Available at: https://www.aua.gr/ns/project/hortimed/Deliverable_2.pdf (accessed on 20 August 2020).
8. Roy, S., Vyas, R., & Kumar, K., (2015). Management of micro-irrigation system in arid areas of India: A conceptual framework (10[th]edn.) [ebook] International Journal of Applied Engineering Research, pp. 3689–3693. Available at: https://www.researchgate.net/publication/280086801_Management_of_Micro_Irrigation_System_in_Arid_Areas_of_India_A_Conceptual_Framework (accessed on 20 August 2020).

9. Seminis, (2019). Water Management Using Drip Irrigation | Seminis. [online] Available at: https://www.seminis-us.com/water-management-using-drip-irrigation/#drip-system-designs-and-layouts (accessed on 20 August 2020).

Chapter 5

Micro-Irrigation-Technological Advances and Its Application

CONTENTS

This chapter explains about the several numbers of various improvements in technology of micro-irrigation and the application of micro-irrigation. This chapter also explains how to improve the water productivity for yield. This chapter also provide highlights on how it impacts the economics of irrigation.

This chapter also addresses the role of micro-irrigation for small farmers all across the globe.

This chapter also emphasises on the use of micro-irrigation for agriculture, impacts of micro-irrigation on crop yield and water saving, environmentally effect of micro-irrigation on soil and water resources, as well as the application of fertilizers by micro-irrigation.

5.1. INTRODUCTION

In the 60s for the various other irrigation methods as the alternative the development of drip irrigation began.

In the arid or semi regions where water is the main element which is present in shortage drip irrigation was viewed as an interesting new method suitable for these areas at that time.

High income intensive crops which are namely vegetables, flowers, and certain orchards during the 70s era this method of drip irrigation proved to be advantageous especially regarding these crops.

The main advantage is water conservation and this method of drip irrigation seems most suitable for modern farmers in order to conserve water.

Further, apart from the main advantage of water conservation there are also some more additional advantages which basically

Help to improve the crop quality and increase yields and these additional advantages were not usually found.

Mainly in regard to the greenhouses and to crops which are grown under plastic mulching these are the areas in which drip irrigation plays a vital role. Field crops such as pineapple, sugarcane, cotton are the ones which became more established during the 80s by the means of drip irrigation (Figure 5.1).

Figure 5.1. An illustration of micro-irrigation.

Source: Image by Flickr.

Also, drip irrigation helped the establishment broadly into the orchards which mainly include deciduous fruit trees, bananas, vineyards, and citrus orchards, the accumulation of know-how on operating the system resulted in the development of highly reliable drip irrigation systems over the last decades. There are various other advantages of these systems as follow:

1. As compared to the other irrigation methods increase in yield was observed by the drip irrigation;
2. Plant heath turned out to be better and also an improved quality of produce was observed; and
3. By combining water and fertilizers which are basically applied to the medium where there is an optimum water ratio which helps to produce optimal growing conditions in both soil and soil less media and this produce is especially the result of the efficient irrigation which also conserves water apart from these produces.

India and Israel are the two main two developing countries and population in which drip irrigation was introduced during this period of time. Further, as a justifiable irrigation this method drip irrigation was established by certain developments during the 90s era.

1. In rainy region, mainly the drip irrigation could be used as a supplementary irrigation as this method of drip irrigation is itself a unique one. Although the method of drip irrigation is spread around the entire globe where the average rainfall is sufficient for supplying the plant's water requirement but initially this method of supplementary irrigation was originally developed for Hawaii. The months during which there are shortages of water, could be shown by the distribution of rains over the years and when a closer look was taken.
2. Due to the recognition that drips irrigation is quite simple to use the penetration of drip irrigation to traditionally irrigate plots, especially the smaller plots. The method of drip irrigation is also suitable for the use of farmers who own smaller plots and also helps to improve their crop yield.
3. For mainly the necessity to channel resources to on-far plots, for transferring water from place to place and from season to season also along with large scale drainage, the recognition by international institutes involved in irrigation. Further, it should be ensured that the farmers are acquainted with irrigation issues also it was necessary to pay attention to the irrigation method employed in these areas as just to transfer water to small farm holders is not sufficient.

5.1.1. Micro-Irrigation Technologies

The socio-economic attributes and technical of the micro-irrigation technologies are the aspects by which the micro-irrigation technologies can be broadly categorized. And these two categories of micro-irrigation technologies are namely the commercialized, state of the art micro-irrigation systems and low-cost micro-irrigation technologies.

Along with the innovative farmers some systems are also designed by organizations such as the International Development Enterprises (IDE) apart from them there are many low-cost systems which include the Pep see easy drip technology, bucket, and drum kits, micro sprinklers, and microtube drip systems.

Conventional drip and sprinkler systems are the two most sophisticated, capital intensive systems. Poor farmers are basically the ones who are mainly helped by these low-cost micro-irrigation technologies.

Further, the compatibility with the smallholder farming systems and competitive pricing are also supported by these micro-irrigation technologies. Although the extent of economic gains from investments depends on the type of crop, but the farmers can generally recover their initial investment capital between one and three years.

The profitability of investing in micro-irrigation can be easily maintained by the means of subsidies and options for financing from organizations and the various schemes which the government can start as in case of India various schemes as IDE, the aga khan rural support programme (AKRSP), etc.As this profitability is generally not available for these poor farmers therefore this makes a crucial difference in adoption by poorer farmers. Further, from low cost to conventional systems a progression is often been noted. These progressions were mainly noted by the means of the International Water Management Institute (IWMI's) study of micro-irrigation adoption.

5.1.2. Improving Water Productivity and Yield

Over traditional irrigation practices such as flood irrigation a significant yield improvement is generally resulted by the use of micro-irrigation technologies.

For example, in India (particularly in its Gujarat region) the type of the irrigation systems which is used and the crop grown are the two factors on which the improvements from the micro-irrigation technologies dependent and this was proved by the latest researches done there.

The water productivity can be easily increased by drip a system which produces more per unit of water used and this is shown by the various studies for many different crops. The energy which is basically used for the process of irrigation can be saved and it is significant also this energy can be easily saves when water is pumped for this irrigation process (Figure 5.2).

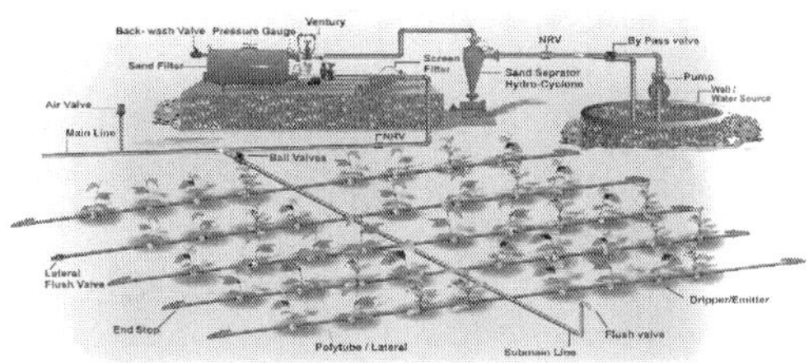

Figure 5.2. An illustration of drip irrigation.

Source: Image by wikimedia commons.

The magnitude of the land productivity differences between the two systems are much less when compared to the magnitude of the water productivity difference between the conventional and micro-irrigation systems. Further, for water scarce the micro-irrigation technologies are more appealing.

5.1.3. Changes in Cropping Patterns

The impact on the cropping patterns in the study area is an interesting outcome of the micro-irrigation. The extent of the cultivation and the crops of the farmers have been changed who has basically adopted the alternative technologies in the study locations.

Further to shift the cropping patterns to high value, water resistant crop and also the farmers are encouraged to increase their overall cropping intensity more specifically by the means of the micro-irrigation adoption.

For example, the major change was observed in case if Maharashtra where the farmers started moving towards the high value crops such as oil seeds, also they also started growing water resistant crops such as banana in place of the traditional crops which were namely groundnuts and oilseeds.

It was also predicted that if farmers move on to the micro-irrigation systems by the studies it will lead to increase in crop yield particularly by the use of drip irrigation.

It was also seen beneficial in the case of crops which were earlier difficult to grow especially when there was early withdrawal of the monsoon and for the crops which were intensified in the same field and also as the system basically allows the growth of multiple crops this resulted in optimum increase in crop yield.

On the lands of the farmers there was an explicit possibility of extending irrigated or cultivated when they mainly adopt the micro-irrigation system. As higher yields bring higher earnings therefore this also led to significant rises in the incomes of the farmers.

As the drip irrigation is a form of precision irrigation and the water is directly applied to the roots of the plants therefore this system has the potential to improve the quality of the harvest. Less amount of water is lost due to the conveyance and also each plant is able to receive the right amount of water at the right time.

It was observed that earlier farmers basically over irrigate or under irrigate their fields when they use the traditional methods of irrigation.

5.1.4. Increasing Incomes and Reducing Poverty

The labor cost reduction, improvement in the crop yield, efficient use of water resources and also the income was also increased for the poor farmers by the use of the low-cost drip systems. Within the monsoon climate there is a particular niche being supported by the means of the drip irrigation system.

This system somehow also improves or it can be said that it promotes the efficient use of the rainwater as by the help of this system the farmers can plant the crops basically it is already established at the onset of rain.

Further the dry spell or the early withdrawal of rain could not affect the crops as they are planted early. Moreover, it also to avoid or avert the crop loss and the decline in the crop yield.

The risk of losing a crop is avoided also one of the most important advantages when basically such type of technologies is being adopted by the farmers that they help to extend the use of water during times of drought or water scarcity.

Therefore, considering the above statement proves that the livelihood security of the poor farmer could be improved and this will in turn lead to the poor farmers who are weak towards the changeability of the rain and deal with them by the help of the micro-irrigation.

As the drip irrigation system reduces the labor apart from the improved quality and yield this further encourages the farmers to adopt this system. The labor cost of the farmers is reduced as micro-irrigation allows for early harvesting of the crops.

Further cutting down the efforts which are made by the farmers which farmers early had to make in order to protect their crops nowadays are done by the means of this system. Also, it was observed that the weeds, insects, and the other plant diseases have also proven to occur less frequently as earlier when the traditional methods of irrigation were under practice.

One of the major reasons why farmers are moving on to this technology of irrigation is that the lower energy cost expended as an overall effect on the cost of production. Thus, reducing the cost and increasing the income of the farmers makes this technique suitable for the farmers.

The labor requirement reduced drastically from using drip irrigation over the flood irrigation which was a traditional method and this knowledge has been come across by the means of the interviews conducted on various places (Shah and Keller, 2002).

By enhancing the household food and nutritional security the drips kits basically provide women with opportunities to earn incomes. As the systems adopted are different so the effect or the impact on rural women differs by adopting the micro-irrigation depending on whether these are small cultivators or the large cultivators.

The women from the large cultivator's household basically used the customized system where as the women from small household cultivator households used drip kits for the vegetable farming in their homestead and this mainly clear by the IWMI's study.

They received revenue from the sale of their produce as it basically provides the women smallholders with an income opportunity. As this technology helps to reduce the labor requirement this, in turn, helps the women from larger farms benefited by it.

The small cultivator households have adopted this technology and they can receive tremendous improvement in the food and the nutritional security by the means of the micro-irrigation technology.

Further in case of the women small cultivators which adapted to this technology of drip irrigation in the homestead vegetable cultivation helped to achieve a marked improvement within the household food security and the nutritional security.

Often using the land that had been bare the vegetables can be grown for the household diets which were otherwise missing from the daily diets and this can be mainly done by the adoption of the drip and irrigation.

The surplus produce which are being sold in the market and this creates further additional income which also helps to improve diets. The women are the ones who are basically responsible for bringing the income in and these additional incomes are remained mostly in the hands of these women.

As women prioritize spending on the household items it has been observed that this particular factor completely lays an impact on the food security of the family (Table 5.1).

Table 5.1. Yield Increases and Water Saving under Drip Irrigation*

Crop	Yield (tons/ha)			Water Use (mm)		
	Conventional	Drip	% Yield	Conven-tional	Drip	Water Saving (%)
Banana	57.5	87.5	52	1760	970	45

Grapes	26.4	32.5	23	532	278	48
Sweet Lime	100	150	50	1660	640	61
Pome-granate	55	109	98	1440	785	45
Papaya	13.4	23.5	75	228	74	68
Tomato	32.0	48.0	50	300	184	39
Water-melon	24.0	45.0	88	330	210	36
Okra	15.3	17.7	16	54	32	40
Cabbage	19.6	20.0	2	66	27	60
Chili peppers	4.2	6.1	44	110	42	62
Sweet potato	4.2	5.9	39	63	25	60
Beetroot	46	49	7	86	18	79
Radish	70.0	72	2	46	11	77
Sugar Cane	128	170	33	2150	940	56
Cotton	2.3	2.9	26	90	42	53

* Full or supplementary irrigation

Source: Table from semanticscholar.org

5.2. IMPACT ON ECONOMICS OF IRRIGATION

The technology also has an advantageous effect on the economics of irrigation as it has a positive effect on the pumping hours needed in order to irrigate a field. This provide assistance in saving a lot of cost as well as energy. In addition to this, it also provides assistance to lessen the wear and tear of the pump and elevates its longevity.

This is permitted with the help of decrement in the quality of water for irrigating a field because of the execution of this technology and this is perceived with the help of both adopters as well as non-adopters.

Nevertheless, the cost of the economics might work out but the conservation of resources can take a hit as these savings in irrigation also provide assistance to expand the irrigated area, in this way, elevating the total water quantity used to earlier levels and also more energy will be needed in order to irrigate the expanded area.

In this way, some of the advantages of drip irrigation with respect to irrigation economics can be lost as well for better farm economies.

Nevertheless, in order to irrigate this expanded area without the drip irrigation would lead in a significant increment in the draught of water which could have very detrimental environmental effects. Drip irrigation provides assistance in order to evade these kinds of extremes (Figure 5.3).

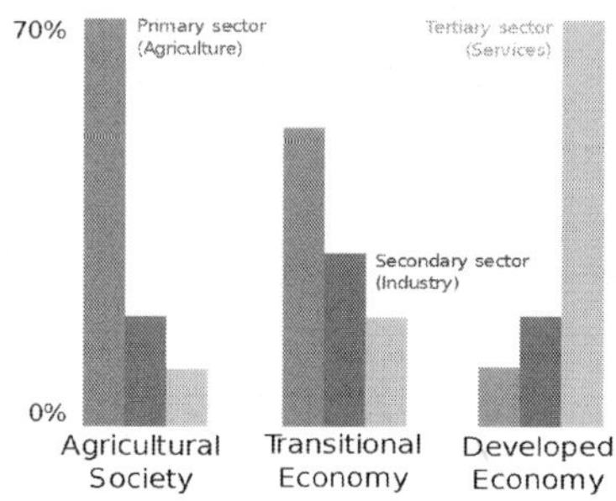

Figure 5.3. Transition from agriculture to developed economy.

Source: Image by Wikimedia commons.

The economics of irrigation is also positively affected with the help of the technology in several numbers of other indirect ways as it makes sure the sufficient and timely water supply to the farm this helps take away most of the variability of farming and productivity, in this way, assuring the productivity and yield and revenues to the farmer.

The assurance of yield and total production allows a farmer to enter into long term contracts and exercise greater market power as well. The technology also has a positive effect on the transaction costs of irrigation as it provides assistance in order to decrease the conflicts and increment in the adaptability in irrigation.

Never the less, the true economics of irrigation will be advantageous only once the scale of adoption in space and time is significant to convert savings into inter-farm and inter-sectorial transfers even if only at the village or household level.

5.3. MICRO-IRRIGATION FOR SMALL FARMERS: WORLDWIDE

More than a hundred studies were listed in commonwealth agricultural Bureau (CAB) documents through the course of the time interval of 1998 to 1999, implemented in Asia, Africa, South, and Central America and on other several numbers of different islands.

1. The Third Conference of On-Farm Irrigation held in Egypt in 1999, distributed mainly with the irrigation development, drip irrigation in desert regions and several other crops. The studies also come up with the usage of saline water as well as recycled water.

The application of drip irrigation is expanding among the conventional small farmers who live in Egypt, and demonstration plots have been assembled in several numbers of various areas. In the desert region of Nubariya, drip irrigation is penetrating the conventional farms in an Israeli-Egyptian joint project, where the development is anticipated to cover area more than 100,000 hectares.

2. Vegetables and fruit: There are an identical development of 12,000 hectares in the other regions of this desert, with a population of 20,000 families, with the leading group that is made up with the help of graduates for two universities training the farmers.

More than four years to five years, the families of these farmers received food and staples from several numbers of different organizations until harvesting their first crops. Before this, they have prepared the soil and installed the drip irrigation systems.

An additional 500,000 hectares have been planned in the other region for enhancing to drip and sprinkler irrigation, and will take nearly fifteen years in order to implement. The Egyptian government gives houses, tools, and subsidies to farmers moving here from the other area.

3. In Nicaragua, in the Ocotal region, the World Relief Organization (WRO)-EME-developed near about a hundred model farms for introducing drip irrigation in the vegetable crops.
4. In Southern Argentina, near about twenty model farms were developed with the system of drip irrigation for vegetables, flowers, and orchards.
5. The use of special drip irrigation systems for small farmers started spreading in Kenya three years to four years ago. These farmers

reported a significant increment in the yields and in income. A group of near about one-hundred and fifty women started implementing drip systems, and within a short interval, this grew to three hundred and fifty women.

It became known that the CEO of the Central Bank of Kenya saw "Family Drip Systems" in use in Malawi, and brought some to Kenya. Because of the success of these systems in Kenya, apart from the other neighboring nations, which is consisting of Uganda and Sub- Sahara nations, also became interested.

6. Talking about Jordan, mainly in the Jordan Rift Valley, near about 64% of the farmers uses the drip irrigation technology. Inappropriate knowledge and poor management have inhibited full exploitation of drip irrigational potential.

The development of the drip irrigation and the likelihood of water conservation are a very serious problem in Jordan, which has suffered from drought for numerous continuous seasons, which produces a decrement of near about 40% to 50% in water available for agriculture.

7. With respect to Nepal, in the hilly areas there are issues with conventional irrigation, but there is a benefits of drip irrigation for small plots. In various numbers of regions of Nepal, observations are being carried out, financed with the help of the Dutch and German organization.
8. In the hilly areas near Lake Titicaca, on the border between Peru and Bolivia, an interesting combination was made between simple rural greenhouses and drip irrigation for growing potatoes and the vegetables all around the year.

The ideology is to use greenhouses in order to safeguards the crops from frost and drought. Potatoes that are grown in the greenhouses produced three hundred and fifty kilograms per fifty-meter square, which is an increment of near about 500% in yield in comparison with regularly grown potatoes.

Following these experiments, Peruvian, and Bolivian researchers are currently experimenting with the growing herbs and medicinal plants which are identical to the greenhouses for the international market.

9. Apart from this, there are identical activities for developing the small plots of vegetables and orchards that are irrigated with the help of the drip irrigation in hilly regions in central Thailand. Here too, it was found that with the help of drip irrigation, there is probability to receive excellent yields when cultivating the small plots.

There are similar activities for establishing small plots of vegetables and orchards irrigated with drip irrigation in mountainous regions in central Thailand. Here too, it was found that with drip irrigation, it is possible to receive excellent yields when cultivating small plots.

Apart from this, other areas where the drip irrigation is penetrating for use in small plots, such as small islands in several numbers of different regions of the world. For several numbers of years there was an issue in numerous islands with the water collection, and agriculture was entirely dependent on rainfall.

10. Over the time period of last ten years, modern irrigation methods were being used to these islands, in order to elevate the level of yields. In the hilly regions, where water sources are restricted, drip irrigation was found to be preferred alternative.

One instance of the effective drip irrigation on the small islands is in sugarcane in Mauritius. There are near about thirty-three thousand small farmers that are cultivating a sum of twenty-five thousand hectares of sugarcane on the island. Most of these farmers cultivate less than 0.5 hectares of land. In the time period of 1970, around 180 small farmers participated in a drip irrigation project.

Apart from this, the other project was started in the year of 1987, experimenting with new diversities with the drip irrigation. It was found that in small plots the production of the sugar was elevated by 75% with the introduction of new varieties with the help of the drip irrigation. The penetration of the modern agricultural methods is still in its first stages on Mauritius.

5.4. USE OF MICRO-IRRIGATION FOR AGRICULTURE

5.4.1. Effect of Micro-Irrigation on Crop Yield and Water Saving

The water requirements in the micro-irrigation can be smaller when it is compared to the other irrigation methods.

And the reason behind this is decreased wetted area; less water is lost to the evaporation. These kinds of systems are also almost no surface runoff.

Micro-irrigation offers a continuous supply of water in the crop zone and has been proven to offer a higher crop yield and increased efficiency of water use over the traditional irrigation method. Postel (2000) claims that drip irrigation has the potential to last double crop yield per unit water in several numbers of applications, which is consisting of irrigation of most of the vegetables, cotton, sugar cane, and orchard and vineyard crops.

A collection of research results from several numbers of Indian research institutes shows that the typical water use decrement with the help of 30% to 60% and the typical yield increment of 20% to 50% for a variety crops, which is consisting of cotton, sugarcane, grapes, tomatoes, and bananas (Kooij et al., 2013).

According to the reports of Gleick (2002) which states that "Shifting from traditional surface irrigation to drip irrigation in India has elevated the overall productivity of the water by near about 42% to two hundred and 55% for crops as diverse as banana, cotton, sugar cane, and sweet potato" (Figure 5.4).

Figure 5.4. Effect of micro-irrigation on crop yield.

Source: Image by Wikimedia

Because of its high WUE, the micro-irrigation is more and more being used as a strategy in order to address the shortage of the water and poverty. Researchers, for example, Shah (2011), have pointed to the water savings and yield elevated because of the micro-irrigation.

Moreover, to the farm productivity (crop yield and output), the income of the farmer and the security of food are also increased. With earlier harvests, the cost of labor is decreased. Enhancements in the drip irrigated crop quality have also been noticed (Maadramootoo and Rigby, 1991).

Talking about Turkey, a well-known researcher Tekinel et al. (1989) directed an experiment which is based on the impacts of the drip irrigation methods and typical irrigation method on the yield of tomato, strawberry, banana, and citrus. The outcomes depicted that the highest yields and crop quality were accomplished with the help of the drip irrigation.

In addition to this, Cetin (1996) directed an experiment with the help of a surface (furrow) method of irrigation on second-crop maize in Harran Plain of Turkey, and attained the highest yield of one hundred and fifty kilograms per hectare, and determined the requirement of water for the irrigation of maize as 1303 mm.

A well-known researcher named as Değirmenci et al. (1998) conducted a research in the same plain with the help of the furrow irrigation and they obtained an average grain yield of two hundred and sixty kilogram per hectare and 873 mm of irrigation water.

On the other hand, another well-known researcher named as Yazar et al. (2002) obtained the highest yield of nine hundred and twenty kilograms per hectare and on second-crop with the help of drip irrigation and applied 581 mm of the irrigation water.

If these values are being compared to the observations, then the drip irrigation could save by as much as 55% when it compared with the surface irrigation. On the contrary, a grain yield of 15% to 23% would be possibly higher (Cetin, 2004)

Several studies on cotton in the Harran Plain of Turkey showed that the requirement of the water could be 1148 mm (Karaata, 1985), 1113 mm (Kanber et al., 1991) and 937 mm (Cetin and Bilgel, 2002) by furrow irrigation.

On the other hand, Cetin, and Bilgel (2002) disclosed that the requirement of the water for cotton could be 619 mm when using drip irrigation in order to yield same amount of area. In this way, the drip irrigation resulted in not only higher production of cotton but also significant amount of water savings as well.

The outcomes of the experiments that are conducted by Cetin et al. (2002) depicted that the average fruit production of fresh market tomatoes that were irrigated with the help of drip irrigation was 1322 tons per hectare. On the contrary, the yield was five hundred and forty-eight tons per hectare when the irrigation was done by conventional irrigation methods and local conditions of Eskişehir.

5.4.2. Environmentally Effects of Micro-Irrigation on Soil and Water Resources

Out of many, the other meaningful advantages of the micro-irrigation systems are that water which is having high content of salt can be used by the system. Also treated and untreated wastewater can be applied in the manner which targets only appropriate crops.

Micro-irrigation is more suitable as compared to the other irrigation methods for the reuse of wastewater because there is no aerosol generation and no wastewater come into the contact with the plant foliage. There are also several issues with the odors, ponding, and runoff.

Moreover, several numbers of studies that recommends that the content of nitrogen which is present in the wastewater is better absorbed by the plants and less likely to pollute the groundwater when applied straight to the root of the plants. Through the course of using wastewater with micro-irrigation it is very important to make sure that the emitters do not clog.

Systems must be observed closely in order to make sure a uniform application and full functionality (Casey et al., 1999).

With the help of a good management of the micro-irrigation systems, the root zone moisture content can be maintained near field capacity all across the season that provides a level of water and air balance which is close to the ideal condition for plant in order to grow.

Through the course of the dry season in humid regions, or in arid climates, micro-irrigation can have a considerable impact on the soil salinization.

In addition to this, drainage issues do not take place. In spite these direct, private benefits to the adopters, several other indirect, social benefits of the drip irrigation: it decreases the soil erosion and non-profit pollution, and the reason behind this is that the water percolates only to low depth of soil.

In this way, the fertilizers and pesticide residues do not mix with the water table, it promotes more effective use of nutrients, it also makes sure the better and longer moisture retention in the root zone.

5.4.3. Application of Fertilizers by Micro-Irrigation

Perhaps, the drip irrigation is one of the most efficient methods of water application to the crops. It develops a limited root system which necessitates regular nutrient supply. This may be fulfilled with the help of applying the fertilizers with irrigation water with the help of fertigation.

An important innovation has been the development of the fertilizers injectors which can be attached to the micro-irrigation systems in order to permit the fertigation, in this way, enhancing the crop nutrient management.

Micro-irrigation systems allow for a high level of control of the chemical applications. Both water and fertilizers can be applied all across the growing season in judicious amounts in order to match the requirement of the crop.

In addition to this, the other chemicals, for example, nematicides, herbicides, growth regulators, along with the fungicides can be very effective when applied through micro-irrigation systems in order to enhance the production of the crop.

Moreover, the frequency of pest and weed invasion, and the other plant diseases, take place less frequently.

And the reason behind this is that a reduced wetted area and drier soil surface. With the help of decreasing the labor which is required to safeguard the crops, and by decreasing the use of pesticides and herbicides, there is an overall financial saving to the farmers (Varma et al., 2006).

Teixeira et al. (2011) stated that the fertigation encourages an increment of near about 36% in nutrient use efficiency when it is compared to the typical fertilization, for either nitrogen or potassium in banana crops.

Moreover, to this, several numbers of studies that are conducted by Darwish et al. (2002) depicted that the fertigation with the help of the drip enabled higher nutrient efficiency when it is compared to the conventional fertilizing.

5.5. TECHNOLOGICAL ADVANCES IN MICRO-IRRIGATION

Technologies of micro-irrigation are more and more seen as a means of addressing the increasing competition for rare water resources. Suitable low-cost drip systems have shown to have beneficial impacts on the yield, incomes, and security of food.

With the appropriate institutional support, these kinds of systems can provide assistance to the poor farmers in order to enhance the water efficiency and income.

5.6. BEST DRIP IRRIGATION TECHNOLOGY TO INCREASE PRODUCTION CAPACITY

Manufacturers will need to stay aware of these trends as the new technologies become increasingly common. Further, the manufacturers will also commit to these technologies as they want to upgrade their equipment and also want to keep up good relations with their customers.

There is an explicit need to operate the drip irrigation pipe with the smart systems also it must be produced in an efficient manner in order to achieve the desired results. One the other hand there are certain challenges which should be overcome so these challenges are not only for existing for the customers but also for the manufacturers.

It could be quite challenging for the manufacturers to explain the working of the system to their customers and how this technology could be helpful for their customers in reducing the cost. Also, there are various challenges about the advantages of new technology (Figure 5.5).

Figure 5.5. An illustration of water sprinkler: A method uses in micro-irrigation.

Source: Image by Pixabay.

The key to operation of the system basically depends upon both the cost-effective and environmentally friendly so the major components which act on these two operational principals explicitly depend upon the increasing production of the crops while minimizing the water use.

Further, there is a need to optimize the approach related to the irrigation system as the fact is that water is a critical resource and its preservation requires an irrigation system.

The groundwater levels, the current growth stage of the plants, forecasted weather, the climate of the area, the current and many more are factors on which the smart irrigation system design must take into account apart from just the plants being treated and the fertilizers which the plants require for their optimum growth.

The manufacturers who are looking for ways to reduce the cost of the raw material for them minimizing the resources are also a matter for concern. Also, these manufacturers want to produce the same level of product by reducing the energy usage.

Further by combining the different polymers in their hose and using more efficient machinery these advances in the dripping technology basically allow the manufacturers to meet their goals. The operational cost is kept reasonable by these advances which are mentioned above also these advances allow manufacturers to meet increased demands for drip irrigation.

5.7. CHALLENGES FOR ADOPTING MICRO-IRRIGATION TECHNOLOGY

Figure 5.6. Challenges for adopting micro-irrigation technology.

Source: Image by the blue diamond gallery

1. On monsoon there still continuation to be gamble for irrigation. The surface runoff which is basically caused by the means of precipitation further results in the problem of soil erosion. It has been further recorded that only 18% of the precipitation; (Figure 5.6).

2. In order to justify the huge investments on high cost irrigation projects there is an explicit need to check on to various things such as checking on to soil erosion various methods, and the efficiency of the utilization of the conserved water/harvested also these requirements should be integrated within;
3. A vertical growth has been brought up basically by these micro-irrigation technologies. Further, these technologies are also considered as rewarding for the purpose of irrigation due to their explicit advantages;
4. They can work excellently in a baffling situation for the farmers. Further, there are also many reasons due to which these micro-irrigation technologies have remained limited with slow progress and slow their adoption has remained slow and all such reasons or factors are mentioned below:
5. The micro-irrigation system is generally considered out of reach of the poor farmers as the initial cost of the systems is high;
6. Also, this micro-irrigation system is generally viewed in isolation because this technology was not integrated with farm irrigation management;
7. The infrastructure for training of farmers, appropriate technical know-how, trained human resources and adequate credit facilities were lacking;
8 The acceptance of this micro-irrigation system by farmers needed much persuasion as this technology is generally perceived as technology-intensive; and
9. There was a lack of availability of low-cost water-soluble fertilizers and other agrochemicals, fertigation scheduling, location-specific and crop-specific irrigation, the optimal; fraction of soil to be wetted, also a lack of information on temporal and spatial variation in soil moisture.

5.8. FACTORS INFLUENCING THE ADOPTION OF MICRO-IRRIGATION TECHNOLOGIES

In the analyses of the results of the logit adoption model, there is a need to probe into the mechanics of the hypothesized relationships between the dependent and the explanatory variables. Most of the expected signs and most of the variables were mainly included in the model.

The factors which were included to indicate the status of the labor availability in the household are basically the family size and the dependency ratio (i.e., the proportion of the family members whose ages are less than 14 or more than 65).

Also, these factors show an explicit insignificant effect reflecting the lower labor requirement of micro-irrigation technologies especially when they are basically compared with the traditional irrigation methods.

Due to the fact that these technologies need technical and the managerial skill for proper utilization for the human capital variables are of special interest in relation to the micro-irrigation adoption process. Further in case of the adoption probability in both Gujarat and Maharashtra the age variable has a positive but an insignificant effect on them.

A lower chance of adoption of new innovations normally one would explicitly expect the older farmers attracted towards these technologies as they only prefer to use their traditional methods of irrigation (Neil and Lee, 2001).

Further, it can be concluded or inferred by the means of the above context that the factor of the age in case of irrigation methods actually shows its real field effect. Also due to the age variable may have been captured more the experience and wealth aspects than its conservatism dimensions the observed unexpected sign of age in the present analysis.

The statistical significance only in the case of place name Gujarat in India shows the interaction between the education and the age expected sign.

Further due to the micro-irrigation systems have a long history and are no more considered as the new technologies in the areas basically shows the insignificance of these variables in many other places of the world. Micro-irrigation technologies can be used by most of the farmers in the other words.

5.9. PROBABLE LIMITATIONS IN MICRO-IRRIGATION TECHNOLOGIES

The technology of micro-irrigation is still quite expansive for the farmers, although the cost of micro-irrigation is reduced by the technological development and the simplification. Further, for the maintenance, optimal efficiency and mostly importantly for the proper system design there is an explicit need for a high level of technical skill.

This technology increasingly attractive to the farmer's various other things could fulfill the purpose these are basically advancements to further reduce costs, labor, and to facilitate the initial system design and installation.

In the developing countries, small farmers have been slow to adopt this practice mainly due to the high initial investment required for taking up or switching to micro-irrigation systems. Further, there are various reasons due to which further reduces the desirability (Figure 5.7).

Figure 5.7. Probable limitation in micro-irrigation technology.

Source: Image by Picpedia.

These factors are the lack of cash and expertise and the crop specificity (Varma et al., 2006). Particularly, with respect to the water quality improvement, filtration techniques and reduced emitter clogging there is basically a potential to capitalize on the use of nanotechnology and biotechnology in micro-irrigation.

By the means of chemical precipitates, organic matter, algae, bacterial slime are the particles which could lead to easily clogging of the small openings.

Further, this system basically needs very good filtration just for the above-mentioned reasons. With the potential to operate on the side, in real time it becomes easy to recognize the nano material-based biosensors, and also to monitor and to measure the presence of the contaminants.

A high level of accuracy, rapid response rates, compact sizes and robustness are achieved by the means of Nanotechnology which is used in soil moisture sensor designs. In transferring research into practices along with the innovation is the main challenge (OECD, 2011).

In low quality water and in saline water can be used in micro-irrigation. Further, during the prolonged dry periods a problem can occur from the salt which is accumulated at the edges of the wetted zone. Injury to the plants could be caused by the means of these salts when they went into the root zones by means of light rain.

To leach accumulate salts from the soil between the growing seasons there is need of an additional irrigation system (sprinkler or surface) (Cetin, 2004). From all topographic conditions from humid to arid and semi-arid regions these are the areas where the landscape applications are applied and also it is observed that the in horticultural and ornamental micro-irrigation has been particularly successful.

Further the elimination of wood and diseases, reduced labor, improved fertilizer application were basically well recognized and these were the advantages of the micro-irrigation system with respect to the water and energy savings.

Also, to expand acreage under micro-irrigation factors such as advances in emitter and dripper technology, the introduction of inexpensive drip tape, and the development of low cost of low-cost sand and screen filters helped out a lot.

Mainly in developing countries for the production of cereal crops a significant challenge is to apply drip irrigation technology. There are many challenges which need to be overcome and these also include challenges such as several social, technical, and institutional challenges. Further, a transfer must be accelerated by education and knowledge.

In terms of conservation or sustainability management of soil and water resources, the most vital or most important is basically a result of using the micro-irrigation systems. Moreover, the micro-irrigation system is praiseworthy technology and also it is highly relevant.

This system is considered as a leveraging technology to sustainable agriculture as the micro-irrigation system is an intervention to address various issues of agricultural growth. If the economic return is more then only the farmers will adopt this technology of micro-irrigation system.

Moreover, to keep water productivity more relevant in economic criteria the economic considerations can be incorporated with more engineering approaches.

By the end of 2026, it is predicted that the global micro irrigation system market will nearly project to reach out to USD 23.4 billion and during a period of 2019 – 2026 the growing CAGR at 12.2 per cent.

In the field of irrigation various new technologies which have been evolved while the period of modernization. The crop irrigation methods have been revolutionized and it is widely accepted and implemented across various countries and all of this is possible by the means of micro-irrigation technology.

Further, the water is supplied in a different case of the micro-irrigation system when compared to the traditional methods of irrigation in this the water is supplied below and above the soil discrete drops, continuous or tiny streams through emitters connected to the water supply.

Drip irrigation and sprinklers are the two major types of the micro-irrigation technology. In sprinkle irrigation, the water is sprayed to the upper surface of the soil using spray heads which is like natural rainfall. In this drip irrigation the water is supplied to the roots of the plants through small diameter pipes in water flows at very small rates.

5.10. SUGGESTIONS FOR BETTER ADOPTION OF MICRO-IRRIGATION SYSTEMS

In order to take the complete advantages of the benefits which are basically offered by the means of micro-irrigation technologies? Also, there should be large scale adoption of these technologies and also, they should be encouraged as they act as the key government policy.

Further, the system of micro-irrigation should not be seen as the end.

Micro-irrigation needs to be promoted in a holistic manner with an end to end approach with integration of agronomic practices and linkages for marketing. Also, it should not be viewed as end in itself.

As these micro-irrigation technologies help in improving the individual and combined efficiency of all other agricultural inputs such as fertilizers, energy, plant protection measures, and due to these micro-irrigation technologies should be viewed as total plant support systems.

As the upstream population as well as population beyond the tail end of the canal have been deprived of the project benefits this is the reason it is accepted as the major and medium irrigation projects have created islands of prosperity amidst surrounding poverty.

Even once the natural share in the water flowing through the river has been received by the deprived lot has also lost. The farmers resort to cost-intensive lift irrigation schemes as they have no other option basically on upstream and beyond the tail end.

Further, there could be in prohibitive water and energy costs rendering the schemes unconventional particularly when there is basically inefficient use of water through conventional methods results in the above-mentioned conditions.

Larger benefits to greater number of farmers with same volume of water is ensured if the government basically adopts the policy of making MIT mandatory specially for all the commercial crops including both the systems the canal and lift irrigation system. A greater equity will be leaded by the means of the policies.

The method which is being used commonly among the developing countries is the drip irrigation system also the only source of water available in these areas is the groundwater. Further by allotting certain proportions of water from each irrigation project only for the use by micro-irrigation.

Along with various technologies and by the means of the knowledge micro-irrigation is leading agriculture. Further to attain sustained and enhanced agricultural production and productivity with the farmers as the center point, they are viewed as one of the eco technological approaches.

Further, fertilizers, and manures and weed removal, improvement in the water use of efficiency, reduction in cost of water, increase in yield are basically some of the imported noted benefits which are approached by the means of the micro-irrigation system.

Such irrigational benefits are turning out be exceptionally outstanding benefits for the farmers. Further, as the there is an optimum utilization of water all these added up in the increase in the overall economic benefits accrued.

Three major components of the Green Revolution are the improved seeds, water, and the fertilizers. And of these three components are bounded together to maximize the synergic interactions by the means of this technology.

To ensure the congruence of sustainability, productivity, profitability, and equity an evergreen revolution is achieved by the green revolution transformation and this also possible due to the means of the micro-irrigation technology.

Also, it has been observed that this technique is entirely relevant and praiseworthy. The evergreen revolution could be achieved without the burden of environmental degradation as the micro-irrigation greatly enhances water, fertilizers, and the energy use efficiency and the promotion of precision agriculture.

Also, with the more engineering approaches to keep water productivity more relevant in economic criteria, therefore the economic considerations can be incorporated.

5.11. CONCLUSION

Over a period of time, it has been observed that within the irrigation patterns the efficiency has been enhanced as this concept of irrigation is as old as the human civilization. Further, as an important need of the day is the adoption of the improved water management practices.

Moreover, the present methods of the irrigation are basically proving to be in vague as they are leading to wastage of huge amount of water. Water management practices such as adoption of water saving methods such as micro-irrigation is considered important step in the agricultural production apart from all the beginnings of the high yielding varieties.

In many countries of the world, it has been observed that the micro-irrigation is working as a revolutionized agriculture. The volume application of this irrigation system is of the water directly to the root zone or on the land surface beneath the plant. Also, this process of micro-irrigation us slow. Further, the moisture if the soil is nearly maintained to an optimum level by this system, and the main or the fundamental concept of this system is only at the root zone.

REFERENCES

1. Dr. Khadeeja, P., &Ratansharan, P., (2017). Micro-Irrigation: An Efficient Technology for India's Sustainable Agricultural Growth (1stedn.). [ebook] Vallabh Vidyanagar, India: Kalpa Publications in Civil Engineering, pp. 398–402. Available at: https://pdfs.semanticscholar.org/4452/5c3830847b57b1fe6deb61810f2b35f85662.pdf (accessed on 20 August 2020).
2. Dr. Uriel O. R., (n.d.). Technological Development of Micro-Irrigation and its Applications in Developing Countries. [ebook] pp. 1–8. Available at: https://pdfs.semanticscholar.org/4e10/5e3c42525c76bf20789f7e4a1021b47f5107.pdf (accessed on 20 August 2020).
3. Drip Irrigation Industry, (2017). The Best Drip Irrigation Technology to Increase Production Capacity. [online] DRTS-Manufacturing Solutions for Drip Irrigation, Pipe Extrusion, and Mining. Available at: https://drts.com/the-best-drip-irrigation-technology-to-increase-production-capacity/ (accessed on 20 August 2020).
4. Neşe, Ü., Öner Ç., & Murat, K., (2013). The Role of Micro-Irrigation for Modern Agriculture. [online] Researchgate.net. Available at: https://www.researchgate.net/publication/259744265_Micro_Irrigation_for_Modern_Agriculture (accessed on 20 August 2020).
5. Regassa, E. N., Bhawana, U., & Nagar, R. K., (2005). Adoption and Impacts of Micro-Irrigation Technologies Empirical Result from Selected Localities of Maharashtra and Gujarat States of India. [ebook] Colombo, Sri Lanka: International Water Management Institute. Available at: https://www.iwmi.cgiar.org/Publications/IWMI_Research_Reports/PDF/pub093/RR93.pdf (accessed on 20 August 2020).
6. Shilp, V., Regassa, E. N., &Samyuktha, V., (2006). Promoting Micro-Irrigation Technologies That Reduce Poverty. [ebook] Colombo, Sri Lanka: Water Policy Briefing International Water Management Institute. Available at: https://dlc.dlib.indiana.edu/dlc/bitstream/handle/10535/4682/WPB23.pdf?sequence=1 (accessed on 20 August 2020).
7. Shiv, K. S., & Mayank, S., (n.d.). Micro-Irrigation an Innovative Technology- its Importance, Challenges and Present Scenario in India. [ebook] pp. 2–9. Available at: https://www.icid.org/9imic2019_papers/9th_imic_st1_paper%20(13).pdf (accessed on 20 August 2020).

Chapter 6

Potential of Solar Energy in Micro-Irrigation

CONTENTS

In the chapter potential of solar energy in micro-irrigation, the importance of solar energy in micro-irrigation is discussed. The chapter also explains the role of solar power irrigation as a solution to water management in agriculture. This chapter discusses how does a solar irrigation system works and where is solar irrigation system is happening. it also explains seven key benefits of using solar power irrigation system. The concept of green technology is also explained in this chapter.

Design principles and consideration for solar-powered micro-irrigation system are also explained in detail in this chapter. It also highlighted fundamentals of spy micro-sprinkler irrigation system.

The chapter also throw some light on portable solar photovoltaic powered pumping system for micro-irrigation system in vegetable cultivation. In this end, the chapter explains some current and future trends and developments in solar powered irrigation systems.

6.1. SOLAR-POWERED IRRIGATION: A SOLUTION TO WATER MANAGEMENT IN AGRICULTURE

Figure 6.1. Solar powered irrigation system.

Source: Image by the financial express.

Farmers are the one who plays an important role in every society, as food is provided by them to the world's population. Though, they provide food and on the other hand one may should not forget that along with the food they also provide energy, which today is considered as a supreme importance, especially as in light of renewable energies. So, it is clearly said that from the wind farmers can produce energy, the sun or the biomass and farmers can use this energy for their own farm, or, if they have a surplus, resell it to companies (Figure 6.1).

To produce the energy solar energy might be one of the easiest ways for farmers. Indeed, it has been observed that farmers usually have several large buildings whose roofs are directly under the sun, without being hindered by the shadows of the trees, and this helps in turning them into an ideal place to settle a photovoltaic system.

Therefore, it is said that the energy produced from this renewable source can be used either on the farm or in the local power grid and the use of solar energy in agriculture is becoming increasingly popular, it also helps the farmer in the way of providing with an additional income.

Irrigation is considered as one of the areas in agriculture that get benefits the most from solar energy, mainly in arid regions. A virtuous circle is represented and that is why it is consider as the main reason for using the sun for irrigation when the sun shines, as it feeds the irrigation system and it is known by everyone that crops needs more water when the sun shines a lot. Therefore, a large quantity of energy is available when it is actually needed.

6.1.1. How Does a Solar Irrigation System Work?

It has been seen that there are some of the pumps used for the transport of the water are equipped with solar cells. It has been said that through a generator the solar energy absorbed by the cells is then converted into electrical energy and after that which feeds an electric motor driving the pump.

Mainly with a diesel engine or with the local power grid most of the traditional pump systems mainly work. However, it is said that these two modes of operations present disadvantages compared to solar pumps.

The access to the electricity grid is not always guaranteed, especially in developing and emerging countries in many rural areas. So, in this case, farmers cannot rely on the traditional irrigation system. Thus, for the farmer to secure a safe power source and for the public grid to avoid saturation using an independent and alternative energy system can be a solution.

AC powered pumps are considered more efficient than the Diesel pumps, as they slightly allow greater flexibility. However, this system relies on the fuel availability as it is one of the main constraints, which is added to a greater impact on the environment.

Solar-powered pumps are one of the pumps which is more expensive that Diesel-driven pumps but the operating costs are quite high and depend heavily on the diesel price. It works the other way round in the solar-powered systems, although this system isrelatively expensive, and the source of energy is free, therefore, after the paying off period, there are no longer operating costs (only the maintenance costs must be considered). Therefore, solar pumps are considered as the feasible long-term investment.

It has been observed in several studies, such as Water for wealth and food security by Ag Water Solutions Project, have shown, the access to water for agricultural, and the purpose remains critical in some areas such as in arid regions of Africa and Southern Asia.

There are many Indian and African farmers which fetch the water directly from the well or the rivers and irrigate their fields by using buckets. Motorized pump is the pump, it is said that if farmers of those regions could have access to a, they would increase their yield by 300%.

Therefore, nowadays, creating solar pumps that are affordable in arid regions R&D tends to focus on that. With the Siemens the company IBC SOLAR has developed a solution to replace diesel engine by solar-powered engine. So, in this case, the whole irrigation system including the pump can remain as it is; only the diesel engine is replaced by a photovoltaic system and the so-called "IBC pump drive controller."

In 2015 in a farm in Namibia there is a prototype of this system which was tested and it is said that according to the manufacturer it is turned out to be quite efficient. There are low acquisition costs as the existing infrastructure is used and this is the main advantage lies in the fact.

6.1.2. Where Is Solar Irrigation Happening?

Africa, India, and South America are the arid regions in which there is the installation of solar pumps such as is also part of many development projects and the main aim is to increase the productivity of the local farmers and this is done for the purpose of improving their living conditions.

There is one of the successful examples of this is the initiative of a Physics teacher in a school in Blankenese (Germany) where students have developed two solar-powered pumping systems in cooperation with the company SET GmbH from Wedel.

There are two farms in which they installed these systems in Nicaragua to pump underground water. With the collaboration of the UNAN University in León, which deals a lot with the exploitation of solar energy this project could also be achieved. Indeed, over the 10 years the project has been running and in Nicaragua 30 pumps are in operation now.

Nicaraguan company Enicalsa is the one by which it is supervised and that also helps farmers in getting benefit from solar irrigation. To produce all year round the use of solar pumps allows the latter, even in dry season and thus to increase their income and strengthen their position in the local market.

As it is previously mentioned that aside from the regions, there is also an increasing interest in solar irrigation systems in Europe. It has been observed that few months ago, a mobile solar drip irrigation system from Austria has reached the production stage.

The Austrian company Wien Energies carried out this project which pursues a dual objective: on the one hand, reduction of CO_2 emissions owing to the use of solar energy, on the other hand, achievement of 30% water savings thanks to the drip irrigation method versus the traditional sprinkler irrigation.

It is said that the principle of the drip irrigation method is very simple. And this is because of with the use of various valves, hoses, and pipes, water drips slowly and at regular intervals to the roots of the plants.

Therefore, it has been observed that there is no wastage of water, as it goes directly where it should go, contrary to a sprinkler system in which water evaporates into the air or seeps into soils where no plants grow. Therefore, there is a method drip irrigation method which enables to grow more crops with less water, turning it into a highly efficient irrigation method.

For proficient water management in developing countries, the drip irrigation system can come as an efficient source, especially in countries who suffer from high temperature and lack water resources. Solar drip irrigation system can help farmers in many ways.

They can help them save water, money, and energy which are the most important factors for farmers, as they face such circumstances on day to day basis. Many R&D projects are functioning for the democratization for the solar power in field of agriculture as this system comes out to be expensive and difficult to settle. But still, it is being considered that not only for future, even today to prevent food and energy crises solar power in agriculture can come as an important source of management.

6.2. 7 BENEFITS OF HAVING SOLAR POWER IRRIGATION

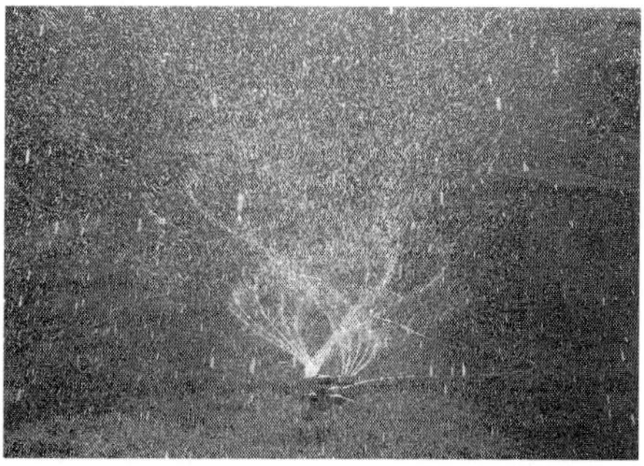

Figure 6.2. Advantage of irrigation to disseminate water to large areas.

Source: Image by Pikist

For many fortunate reasons the use of solar energy is being increased in between farmers for their irrigation needs. As one knows that on large scale farming comes as a very expensive venture so the incorporation of solar irrigation offers some long-term offsets such as channeling certain costs at the right and critical sectors of the farm. Below are seven reasons you need to have solar irrigation installed your farm (Figure 6.2):

6.2.1. Solar Energy Is Environmentally Friendly

Greenhouse gases are the big factor in environment today as it has led to some serious issues in environment endangering ozone layer and other factors therefore, unlike biogas and coal it does not emit greenhouse effects.

It is in turn very environmentally friendly other than just taking care of watering the crops. It also helps in other factors of environment such as reducing noise pollution; it doesn't make much noise unlike generators used for irrigation. Moreover, it helps in saving water as it does not require water for maintenance purpose resulting in saving more water for irrigation of crops rather than maintenance.

6.2.2. Solar Energy Is Free

Once the solar panels are installed on the farm, one does not need to worry about the monthly electricity bills; it is being settled for years to come as the solar energy is free. But one needs to make sure about the quality of solar panels and seasoned expert directions to install it.

6.2.3. Solar Energy Is the Perfect Alternative to Electricity

Electricity cut-offs and unprepared blackouts of electricity can be really difficult especially for farmers using electricity to run their irrigation systems, it could cause major problems as well. Solar energy comes as major effective alternative of generators which in turn are very expensive and difficult to run and maintain. Solar energy is free of cost as it on runs on sunlight so no worrying about system shutdowns or failures as the systems is pretty too straight forward to operate as well.

6.2.4. The Solar Energy Improves Grid Security

Due to penetration of solar energy, the grid is always protected because there are very few blackouts with the network with power plants spread within the wide range. It also helps in unexpected natural calamities. The grid is protected by the power plant resulting in improving security of the grid.

6.2.5. Solar Panels Are Applicable

As the solar energy if free of cost, one can use the energy for various other services in the farm such as lighting the farm. Other than just irrigation it offers wide variety of benefits associated with installation of solar panels. As a bonus for installation one can observe drop in utility bill as well.

6.2.6. Job Creation

Solar panels can help create the jobs and employment in and around the locality. The usage of solar panels can increase business within the locality. One can get all the information on installation of solar panels from local business centers dealing with solar energy for running the irrigation systems and they could also hire people from the same companies for future maintenance resulting in creating more jobs and business for their employees and employers themselves.

6.2.7. Solar Energy Is Efficient

Due to short distance between panels and systems, the solar irrigation system runs more effectively and efficiently. One can be guaranteed about the usage of 100% energy within the farm for the farm only as the solar panels are installed within the farms resulting in no wastage of energy.

Farmer does not have to face any supply cost unlike electricity and can use solar panels for irrigation and other useful purposes and can easily rely on the solar system for its management.

The installation of solar irrigation system can result in many benefits for the farmer who's looking forward for maximizing its harvest. It is one of the most beneficial ways for not only today but for future long-term growth of farming.

It not only helps in cutting down of utility bill but also helps improves its efficiency resulting in better harvest and prevent energy wastage. Moreover, it helps environment in a big way. One will be pleased with its long tern run if put aside its initial installation cost.

6.3. CONCEPT OF GREEN TECHNOLOGY

The green technology is a broad term for more environmentally friendly solutions. Green technology is considered as one of the best possible way that can help in reducing the environmental damages caused by the technologies and the products created for making the life of people easier.

Because of this, it has created its image as environment healing technology that contributed heavily in preserving the environment as well as ecosystem. It is widely believed by masses that green technology plays very important role in enhancing the overall profitability of farmers while at the same time conserving natural resources and reducing environmental degradation. The risk concerned with the poorly designed technologies such as the petroleum products are making the current situation worse.

The "clean" technologies, on the other hand, are famous for providing low risk alternatives. Green technology covers a broad group of materials and methods for producing energy to non-toxic cleaning products.

The mono aim behind this approach has been dramatically vital as people expect a dramatic innovation that can improve their conditions and making their life easier.

The development of alternative technology such as the green technology aimed at benefiting the entire planet by truly protecting the environment. Although it is highly challenging to exactly outline the areas that are covered by green technology, yet green technology plays very critical role in addressing the emerging issues of sustainability because of the recent advent in science and technology.

This technology should be able to cope up with the current needs of society in ways that can provide sustainable and secure future without depleting or damaging natural resources. In short, green technology can be defined is defined as the technology that helps in fulfilling the needs and wants of current generation while at the same time knowing the need of future generation by preserving the nonrenewable resources for them.

In terms of the technology, products can be fully re-used or reclaimed by effectively reducing pollution and waste in the cradle to grave cycle of manufacturing process. The recent advancement in the technology trigger the need for formation of alternative fuels as a new means of energy efficiency and producing energy.

The execution of such innovative technology is very important for agricultural sector in order to ensure the adequate food supply for the continuously increasing population and mitigating the present-day's concern of climate change and global warming.

Irrigation is considered as one of the most energy intensive activity in farming sector. The importance of irrigation of small-holdings is likely to increase in the coming decades, particular in the developing countries, because of steady rising population and because the major proportion of land-holdings are small, especially in Africa and Asia.

Hence, the most feasible options for deriving power through green energy technology are from biomass, solar, and wind energy.

PV pumping systems are believed to be adequately appropriate for energy and water saving methods of irrigation such as sprinkler and drip irrigation because of their benefit in the low-energy range. The consistent rise in the price of fuel price and inconsistent supply of electricity are at present the major constraints for achieving assured irrigation for crop cultivation.

Solar pump therefore is considered as the best alternative currently to the electric motor operated pumps for irrigated crop production primarily for the off-grid rural areas.

6.3.1. Solar Photovoltaic System

It is generally seen that electrical energy is play the central role of all developmental efforts in both the developed as well as developing nations because conventional energy sources are limited in nature and are depleting at a higher rate.

In the last few decades, energy related problems are consistently increasing and are fetching the great interest among environment experts as it involves the ideal use of resources, responsible for the environmental impact due to the emission of pollutants and the consumption of conventional energy resources.

Direct conversion of solar energy to electricity is can be done conventionally through the use of PV cells, which makes use of the PV effect. PV effect depends on engagement of photons, with energy equal to, or more than the band-gap of PV materials. In this, some loss can be easily to avert as it is probable to avoid band-gap limitations by cascading semiconductors of different band-gaps.

PV modules possess the ability to produce electricity directly from light without noise, emissions, or vibration. Sunlight is free, can be used by anyone, but power generation cost is substantially high, although prices are consistently plunging. The density of solar energy is comparatively low as PV modules require the use of larger area for small amounts of energy generation.

The major component in grid connected PV systems is the inverter that transforms DC power generated by PV array into AC power consistent with the power quality and voltage requirement of the utility gird. Silicon solar cells are perhaps considered as the simplest and most extensively used for terrestrial and space applications. The PV system can be perceived as trusted source of electricity production for CO_2 emission reduction and energy resource saving, even if existing technologies are applied. Further the development of efficient and effective solar cells, type, and proportion of material used in the solar cell, and the system design for greater use of recycled material will lessen the greenhouse gas emissions and energy requirement. Net annual CO_2 emission mitigation potential from 1.8 kW solar photovoltaic (SPV) pump at an average solar radiation of 5.5 kWh/m^2/day is about 2085 kg from diesel operated pumps and about 1860 kg from petrol operated pumps. The carbon dioxide emissions diminution potential is greater in the case of diesel substitution in comparison with the petrol substitution. It is primarily because of low efficiency of fuel utilization in the diesel engine pump.

The demand of electricity for irrigation activity is constantly increasing, as the price of an electric powered pump is comparatively low to a diesel engine driven pump. Solar pump can be used as an alternative for small scale irrigated crop production in the off-grid areas of the tropical countries supplied with copious supply of solar energy.

The solar radiation varies from 4.0 to 6.5 kWh/m^2/day and the bright sunshine hours vary from 6 to 9 h/day. There is large size of area that need to be irrigated where most of the areas have no grid connection. Solar PV pumps can play crucial role in irrigating these lands for adequate production of agriculture material and to increase cropping intensity. In regions that have high insolation levels, PV pumping systems were technically suitable for use, environment friendly and were economical to use when compared with the diesel engine driven pumps.

The energy crisis is severe when the irrigation season arrives that can be a threat to economic development of any country. Ensuring far and wide use of solar pumps, government of countries can dramatically reduce the energy demand required for the operation irrigation systems.

Though the initial cost acquiring and implementation of solar pump is relatively higher than a conventional diesel engine operated pump, solar pump has lower maintenance and repair cost, which means over a prolong period of time, it is more cost-effective compared with traditional pump. In addition, a solar pump is pollution free and environment friendly water pumping system.

6.3.2. Solar Water Pumping

It is important to note that nowadays electrical and diesel-powered water pumping systems have been used widely used for irrigation applications. The consistent depletion of conventional energy sources and their subsequent environmental impacts have put an alarming situation all over the world that result in forcing people to move towards choosing solar photo-voltaic pumping system in a sustainable manner. In some parts of the world, farmers are still using hand pumps for irrigation of small patch of land.

Electric and diesel operated have been used mostly for lifting water from bore well, dug well, and other irrigation systems. Lifting of water by hand pump requires lot of effort and is considered as one of the most complicated tasks. In a similar way, non-availability, and limited supply of grid connected electricity in the rural areas forced the people to make use diesel to carry out their day to day farming activities. All this arise the need to look for a reliable source of energy for assured irrigation.

Installation pump sets that works on electricity is not a feasible solution in majority of the locations as the agricultural fields are far away from the electric grid station. In addition to this, the cost of electricity has been in increasing day by day and thus increasing the cost of operation of water pumping.

Further, the overall maintenance needed and the repair cost of electric motor used in water pump is comparatively higher than solar PV water pumping system. When there was limited knowledge on the benefits of on solar PV water pumping system, then for farmers, diesel pumping system is their first choice as it is portable and economical in terms of maintenance costs.

During that time, the cost of diesel was also cheaper. But it contributed to global warming and environmental pollution by releasing a large proportion of CO_2 into the atmosphere. The overall maintenance and repair cost of diesel pump comparatively higher when compared with solar PV water pumping system. Hence, in the current time frame, solar PV water pumping system is considered as the best option for farmers as its pumping cost is cheaper as compared to diesel or electric pump sets.

In addition, the contribution of solar PV water pumping system is less and its maintenance and repair cost are very minimal. It is highly portable as it can be installed at any location according to the needs of the farmers, because solar energy is innumerable and free of cost in the nature.

6.4. DESIGN PRINCIPLES AND CONSIDERATION FOR SOLAR POWERED MICRO-IRRIGATION SYSTEM

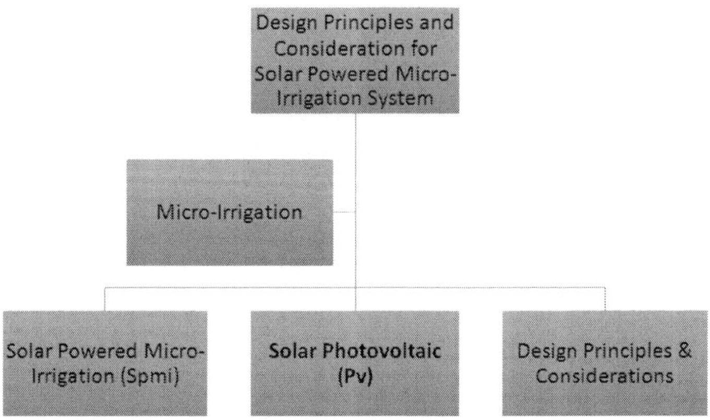

Figure 6.3. Design principle and consideration for solar powered micro-irrigation system.

In broad-spectrum, micro-irrigation can be defined as a pressurized irrigation system that requires gasoline/ electrically powered centrifugal pump-set to produce unswervingly pressure for irrigation of crops largely for almost 16 to 20 hours to produce crop on a day to day basis. This implicates that, while crop irrigation for almost 16 to 20 hours on daily basis, recurring cost of fuel or electricity becomes immense resulting in non-viability and non-sustainability of agro-irrigation projects (Figure 6.3).

Other major challenge faced by the farmers in irrigation system is shortage of power. The gap between demand and supply for power is continuously increasing, putting significant pressure on government on how to deal with this complicated problem in spite of allocation of huge funds.

According to a UN report, it was observed that there is a deficit of 21 to 85% electricity in majority of countries and surging at the rate of 7 to 15% every year. Therefore, majority of attention is given to substitute sustainable and renewable source of energy for instance Solar Power. Although, it is worth noticing that solar radiation is freely available, it has its own drawbacks and not so cost effective to reliable and affordable to everyone.

The concept of Solar power has been widely popular over the past few years all over world in all sectors from space station to agro-irrigation due to its unlimited availability of free solar radiation without non-judicial and proprietorships. Thinking about the use of solar power in micro-irrigation sector is in primeval stage because of lack of information and data on design aspects and remarkable sorts of electro-mechanical machines to pressurized the micro-irrigation system.

According to the world data about the solar power data available and potential for solar energy in various countries outline in Table 6.1, it can be concluded that countries positioned across the equator have a wide scope of harvesting maximum solar power @ 350 watts/m. while on the other hand, Potential of solar energy decreases while proceeding towards north-south poles @ 3 to 5% per degree of latitude.

However, there are various factors on which solar energy harvesting is relied. Some of these are temperature, clouds, relative humidity, longitude, latitude, and obstructions at the specific location.

Therefore, it is essential to acquire the climatic data from nearest meteorological station. On the basis of meteorological data, potential of solar energy harvested can be assessed for design purpose along with solar module and allied equipment.

6.4.1. Micro-Irrigation

Micro-irrigation basically refers to a term that is used for portraying the method of irrigation that can be characterized by the following features:

1. Water is applied at a low rate;
2. Water is applied for a prolong period of time;
3. Water is applied at recurrent intervals levels;
4. Water is directly applied to the root zone of plants; and
5. Water is applied through minimal pressure delivery system.

Micro-irrigation system usually supplied water to the crop by using a piping network and drip lines with drippers spaced at definite intervals along the row of crops is normally driven by diesel engine pump-set, and electrical motor pump-set. Each drippers/emitter supplies a measured, exactly governed by uniform application of water, nutrients, and other required growth substances straight into the root zone of the crops, resulting in triggering the growth of plant enhances and resulted in higher yield.

Table 6.1. Potential of Solar Energy for Various Countries Given for Reference Only

Position	**Countries**	**Solar Power (w/m)**
Near the Equator	Ecuador, Colombia, North Brazil, Gabon, Congo, DRC, Uganda, Kenya, Rwanda, Somalia, Maldives, Indonesia, Tanzania, South Sudan, Ethiopia, Malaysia, Central Africa, Cameroon, Nigeria, Ghana, Liberia, Sierra Leon, Burkina Faso, Senegal, Guyana, Venezuela, Costa Rica.	300–350
Near Tropic of Cancer	Taiwan, Philippines, South China, Vietnam, Thailand, Burma, Bangladesh, India, Srilanka, Nepal, Pakistan, Afghanistan, Iran, Saudi Arabia, Iraq, Israel, Jordon, Syria, Egypt, Algeria, Tunisia, Mexico, Mauritania, Mali, Niger, Chad, Libya, Egypt, UAE, Oman, Myanmar.	250–300
Near Tropic of Capricorn	Botswana, Brazil, Chile, Madagascar, Mozambique, Namibia, Paraguay, and South Africa.	250–300
Above Tropic of Cancer	Japan, North & South Korea, China, Turkey, Russia, Ukraine, Germany, France, Spain, Italy, USA, UK, Other European countries.	200–250
Below Tropic of Capricorn	Australia, New-Zealand, Argentina, Chile	200–250

Source: Image by icid.org

6.4.2. Solar Powered Micro-Irrigation (SPMI)

In micro-irrigation system that are solar power, solar energy (SPV modules) is majorly used to powered motor pump-set unit in comparison to conventional diesel engine or electrical motor pump-set. A typical schematic layout of SPMI system basically comprises of a piping network, control head unit, SPV module, drip lines, motor pump-set, and control panel.

6.4.3. Solar Photovoltaic (Pv)

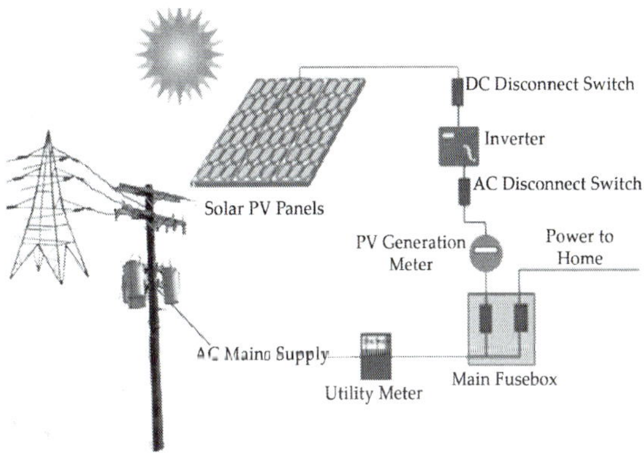

Figure 6.4. Solar PV power plant single line.

Source: Image by Pinterest

The photovoltaic effect can be defined as photons of light exciting electrons into a higher state of energy, letting them to work as charge carriers for an electric current; for instance, solar cells transform sunlight energy directly into flow of electrons. Alexander-Edmond Becquerel (French Physicist) was the first person who observed photovoltaic effect in 1839 (Figure 6.4).

The term photovoltaic mainly refers to the unbiased operating mode of a photodiode in which current via the device is solely due to the transduced light energy. Practically all types of photovoltaic devices are some sort of photodiode. Solar cells generate current directly from sunlight, that can be used to power any equipment or to recharge a battery.

The first practical application of photovoltaic was used to give power to spacecraft and orbiting satellites, but today the major proportion of photovoltaic modules is being used for grid-connected domestic lighting and power generation.

6.4.4. Design Principles and Considerations

Since solar radiation is available at the day time for generation of power, the entire irrigation activity to grow crops occur in a day time to efficiently utilize the generated electricity because it is difficult to restore them and used it later.

Therefore, it is always recommended to use solar electricity immediate after it is produced. Because of this, all the design calculation is based on the fluctuation of solar radiation from morning to evening and from season to season to fulfill the crop water demand. However, some of the factors that should be taken into consideration and integrated while designing micro-irrigation system are discussed as follows.

1. Day time average solar radiation-on seasonal or on a monthly basis;
2. Optimization of ET crop (overall water demand for crop production) on daily, alternate, and monthly basis;
3. Analyzing the water retention capacity on the basis of Soil quality;
4. Master plan of crops;
5. Water assessment and availability to fulfill water demand for crop;
6. Management/Operation of system;
7. Optimization of necessity of motor-pump-set (solar-powered);
8. Techno-commercial parameters; and
9. Photovoltaic Solar Module.

It is important to note that proper planning of crops is imperative while choosing a Solar Powered system. Conventional planning of crops is not adequate and useful and may become costly in overall. Thus, master planning of crops should be in accordance with the overall demands of crops to maximum harvest of solar radiation.

This means, there should not be any scarcity of solar radiation while demand of crop water is at its highest stage. In addition to this, there is need to take some kind of Precautions such as intensification of soil-based retention capacity by adding organic manure and soil conditioner to avert the risk of depleting moisture level beyond the wilting stage of crops even though there is gap of irrigation.

Ensuring proper supply of water is of utmost importance to meet the water demand of crops in the day time when solar radiation is higher.

Design of micro-irrigation components; for instance, filtration system, piping network, and drip lines, etc., should be adequate and corresponding to cope up the demand of crop at varied solar radiations on seasonal basis. In other word, design of micro-irrigation system needs to be synchronized with harvested solar power (radiation) to fulfill the crop water demand.

It is worth noticing that average solar energy can be harvested to 200–300 watt/min several countries base on longitude and latitude. It primarily differs from season to season and mainly affects the haziness and clouds. During the summer season, maximum solar power can be retrieved and minimum during cloudiness such as in rainy season.

Hence, the occurrence of irrigation in rainy season is relatively low and maximum during the time of summer. I

n a similar way, in a day time solar energy is minimum in morning, while maximum in noon and again at its minimum during evening.

Thus, it is right to assume that the micro-irrigation system should be designed in such a way that it can help in meeting the water demand of crop in day time corresponding to available solar radiation only.

In the current time frame, there has been significant research and development is going on to find out the friendly and economical solar powered system across the world at different levels.

Though the overall investment at initial stage is high, considering the recurring cost and dependency such as international price of crude oil and currency fluctuation, government policies, SPMI system is much more economical, sustainable, and deliver promising future for farmers across the world to address the power issues, become self-sufficient and independent in food and power.

6.5. FUNDAMENTALS OF SPV MICRO-SPRINKLER IRRIGATION SYSTEM

Figure 6.5. Spv micro-sprinkler irrigation system.

Source: Image by Water irrigation.

6.5.1. Brief History of Technology

The PV effect was first discovered by a French scientist Becquerel in 1839 who came to a conclusion that more current can be produced if more light is permitted to fall in the cell. He also noticed that the increase in current is primarily relied on the wavelength of light. In the year 1877, Day, and Adam first witnessed the same effect in solids while working with selenium (Figure 6.5).

In the year 1877, Heinrich Hertz revealed that ultra violet light changed the lowest voltage potential of causing a spark to jump between two metal electrodes. In 1905, Albert Einstein described that light behaves similar to a particle rather than a wave. The energy of each light particle, known as photon, is based on its frequency only and is equivalent to the product of Planck's constant (h) and frequency of light (f).

In the year 1954, researchers at Bell Laboratories and RCA, USA narrated accomplishing efficiencies of about 6% by making use of devices made of n and p-type semiconductors. PV cell usually formed from silicon, a material called semiconductor, has now widely been used for production of DC electricity.

Historically, it has been almost 65 years since the operational silicon solar cell was established in 1950s. However, within the last 15 years, there have been significant progresses seem in the solar cell technology that is primarily because of use of silicon.

Over the last many years, there have been improvement in solar cell efficiencies and the cost of solar cell generation has fallen. The efficiency of solar cells present in the market is in between the range of 13 to 16%. Silicon-based solar cell modules are now accessible in the commercial market. There are SPV modules which is constructed from other related technologies, called thin-lm technology, like thin-lm cadmium telluride, thin-lm amorphous silicon etc.

6.5.2. Applications of Spv Systems

1. It can be used for solar street lighting system;
2. In homes, it provides lighting to some areas;
3. Water pumping systems (for drinking water supply and micro-irrigation);
4. Wider applicability for satellites and Space vehicles;
5. It can be used for television sets and community radio;
6. Best suitable for charging of Battery;
7. Plays huge role in monitoring of weather;
8. Power source for navigational lights;
9. Power source for railway signaling equipment; and
10. Power source for various telecommunication equipment.

6.5.3. Advantages of Spv System

1. Absence of moving parts;
2. It possesses the ability to directly transform light to electricity at room temperature level;
3. Can function unattended for prolong period of time;
4. It demands low maintenance cost;
5. It does not cause any kind of environmental pollution;
6. It possesses very long life;
7. Highly reliable and steady in nature;
8. Solar energy is free to use and no fuel is needed for its operations;

9. It can be easily started as no initial time is involved;
10. These have high power-to-weight ratio, which means highly useful for space application;
11. Dispersed or decentralized power generation at the point of power consumption can distribution costs and save power transmission;
12. These can be used with or without sun tracking.

6.5.4. Limitations of Spv System

1. Manufacture of silicon crystals requires high energy and lot of energy;
2. Low efficiency, which means it can run only for minimum time after heavy charge;
3. The insolation is unreliable and thus requires storage batteries;
4. Solar power plants necessitate the need for very large land areas;
5. Cost of electrical generation is relatively higher;
6. The energy consumption in the manufacture of solar cells is very high; and
7. High setup cost.

6.6. PORTABLE SOLAR PHOTOVOLTAIC POWERED PUMPING SYSTEM FOR MICRO-IRRIGATION SYSTEM IN VEGETABLE CULTIVATION

Figure 6.6. Portable solar photovoltaic powered pumping system.

Source: Image by Wikipedia.

The demand of energy has been significantly increasing in every segment of the national developments because of substantial growth and expansion in several sectors such as agriculture, industry, transportation, irrigation, housing, health, communication, education, entertainment, and city modernization, etc. (Figure 6.6).

To cope with this increased demand, in some countries, especially the developing ones, the use of coal-based power plants has been significantly increasing for power generation that pose a serious threat to the environment and is also a major contributor to global warming by releasing a substantial amount of greenhouse gases and hazardous emissions to the atmosphere.

One only available alternative in this context is to provide backup to the existing power sector with non-conventional energy sources. Among the non-conventional energy sources, the solar energy is considered as the most attractive and feasible one as it is available in abundant quality and free to use, mostly available in the tropical areas.

In addition, electricity from solar photovoltaic (SPV) system is gaining more reputation nowadays because of significant plunge in the cost of SPV modules through recent advancement in this area. The major area of focus of policymakers, planners, and researchers is also now diverted to the applications of the SPV system for pumping of water in the irrigation sector because of consistent rise in the water demands in agricultural sector and availability of water has become more crucial than ever before.

In developing countries; for instance, in India, diesel, and electrically powered water pumping systems are used mostly for irrigation systems. It demands significant energy that is limited in countries like India. Developing a grid system is highly expensive as rural villages are normally located too far away from existing grid lines. Even if there is availability of fuel within the country, transportation of that fuel to rural and remote areas is quite challenging.There are no proper infrastructure or road facility to provide much needed supply of fuel in remote areas. Therefore, the use of renewable energy is highly important for water pumping applications in remote areas of many developing countries. While, on the other hand, transportation of renewable energy systems, such as photovoltaic (PV) pumps, is comparatively easier than the other types because they can be conveyed in pieces and then reassembled at original site.

Production of PV energy is considered as core for production of future energy because it is completely free in its availability, produces no pollution or harmful gases, and is highly reliable. These facts make the PV energy resource more attractive for many applications, particularly in remote and rural areas of the developing countries. Solar PV water pumping is widely known for grid-isolated rural locations in places where solar radiation is available in plethora quantity.

Solar PV water pumping systems can supply adequate water suitable for irrigation activities without the need for any kind of fuel or the intensive maintenance as required by electrical and diesel pump sets. They can be installed easily and are durable and highly reliable to operate. It is also modular, comes with latest design that make it easy for future expansion. They can be installed at the site of use, avoiding the spread of infrastructures and long pipelines.

6.6.1. Review of Some Developments in Spv Water Pumping System

The methods for energy security via SPV power system and enhanced WUE (WUE) measures through the use of micro-sprinkler irrigation system in comparison with conventional flood irrigation.

Thus, nowadays a hot topic among the scientists, researchers, and agriculturists not only for ensuring adequate supply of water to the crops, but also preserving the nonrenewable resources and protecting the environment against noise pollution and release of greenhouse gases by the use of rising diesel pump sets.

There have been several attempts put by some researchers in examining the viability of SPV water pumping system for irrigation purposes and domestic drinking water.

Hamidat et al. examined the application of small-scale irrigation with PV water pumping system in Sahara regions. The authors designed a mathematical program to measure the progress of PV arrays under Saharan climatic conditions.

Their research revealed that it is viable to make use of a PV water pumping system for low heads for small scale-irrigation of crops in Algerian Sahara regions. Thus, the PV water pumping system can easily fulfill every day's need of small-scale irrigation with an area smaller than 2 hectares.

They also arrived at a conclusion that the photovoltaic water pumping system (PVWPS) can enhance the living conditions of resultant farmer with the advancement of local farming and thus the migration of rural workforce will be at its minimum level.

Meah et al. also assessed solar photovoltaic pumping system (SPVPS) for remote locations in rural western USA. They came at a conclusion that solar PVWPS (SPVWPS) is an environmentally friendly and at the same time cost-effective way to pump water in remote areas, where 24*7 hours electrical service is not possible and maintenance is also a serious an issue.

From their survey, it was specified that a total of 88 SPVWPS were fitted in all 23 states of the USA, of which 75 systems are still in operation.

They have observed that drought-affected areas such as Montana, Wyoming, Idaho, Oregon, Washington, and parts of Texas can use SPVWPS to enhance their water supply to livestock in remote locations. They were influenced that successful demonstration of these systems is boosting the confidence of other ranchers to use this relatively new technology as another viable water supply option.

They concluded that SPVWPS outperforms other available systems in terms of reliability, productivity, and cost-effectiveness, and the system was able to minimize the carbon dioxide emission substantially over its 25-year life span.

Meah et al. also examined opportunities of SPV water pumping and challenges in the USA. According to their views, it was demonstrated that some improvements can be done to minimize the initial capital investment requirement and to reduce the cost of maintenance and operation services using local-level operation and maintenance.The authors validated that with the help of local resources, such as skills, raw materials, and other valuable equipment, the SPVWPS could be economically viable in developing countries and competitive with the traditional diesel generator water pumping systems. They established that the SPVWPS should be compatible with the local practices and culture to fulfill local needs and desires that can be accomplished by using local resources.

6.7. CURRENT TRENDS AND DEVELOPMENTS IN SOLAR POWERED IRRIGATION SYSTEMS: INNOVATIONS IN TECHNOLOGY AND SERVICES

6.7.1. Solar-Powered Drip Irrigation Systems

When adequately sized, solar pumps can support sprinkler, drip, flood, or pivot irrigation methods. It primarily relied upon the local conditions, a system that can include fertigation or filtration equipment. It is often seen that solar pumps are most of the time combined with low-pressure drip. The necessary pressure is majorly attained by pumping water into a raised water tank and then discharging it with the help of gravitation.

However, the tank offers an extra expense and is generally costlier compared with the pump itself. As flow rate and pressure of a solar pump differ with insolation, the express connection of the drip system to the pump is challenging. Solar trackers play a very crucial role in maintaining power output and thus pressure for a number of hours during the daytime, but these are also highly costly. Putting the solar panels not in north-south direction but divided between east and west directions can also attain the more constant pressure/ power for the irrigation system.Usually, the rate of extra panels will be less costly comparative with the cost of an elevated tank or a tracking system. It is generally seen that irrigation is majorly done on a day to day basis, relied upon the category of soil and crop types as well as the existing climate conditions.

Thus, it can be said that the trade-offs between an oversized pumping system and a storage tank need to be considered. A storage tank that is daily filled by a smaller pump, might be more cost-effective as compared to the larger pump that operate only once a week. The other possible substitute would be to maintain a rotating irrigation schedule.

6.7.2. Floating Solar Systems

Floating solar systems make it possible for the standard PV panels to be fixed on large bodies of water, such as quarry lakes, drinking water reservoirs, tailing ponds and remediation, or irrigation canals. There is no utilization of land resources in this case for the PV panels. Decline in the evaporation of water, higher efficiency of the solar panels and slower algae growth because of the cooling effect of water on the panels provides additional benefits. Easy-to-install and eco-friendly systems are already tested and, on the market, (e.g., Hydrelio, 2017).

6.7.3. Solar-Driven Center Pivot or Lateral Move Irrigation Machines

Till recently, center pivots were (usually) associated with large-scale irrigation. In a present advancement, solar pumps supply water to center pivots. Nonetheless, majority of the present systems still required an external energy source for their overall, functioning, control, and drive units, even if the water expended is being conveyed with the help of solar energy.

Developments are in progress to run the complete tasks on solar energy-likely without any external supply of electricity. Smaller center pivots (approx. 2 to 4 spans = 10–15 ha) are being verified at the moment. Larger systems will definitely come into existence. Solar-powered center pivots (along with batteries) are functioning in field tests beside the Nile. This technology is introduced by the solar and irrigation industry, fulfilling a need of large-scale farmers. Smallholder farmers could manage and assign a center pivot to irrigate nearby land (Hollemann, 2017).

6.7.3.1. Opposite Trends: Integrated, Efficient SPIS vs. SPIS Made up of Components

It is worth noticing that selecting the right type of solar generator, type of pump and its relevant size, as well as technology used in irrigation, is a multifaceted task. In this, the system needs to be fully adapted to the existing site conditions.

Suppliers attempt to supply the entire system, incorporating the solar generator, controller, pump, and accessories, in addition to the irrigation system itself. Moreover, suppliers consistently ensure support of technical (irrigation) in a way to fulfill the need of existing users. It is of utmost important to elevate the use of the system and comes with several benefits. As a result, tenders should ask for estimates for complete systems instead of a single component. In this way, the whole efficiency can be resolute and linked among the submissions.

However, another trend goes in the completely different direction: PV panels, available controllers, typical irrigation pumps present in the market are used and unified into one unit with the irrigation system. This is only applicable for small systems, irrigating up to five hectares (ha), and the overall efficiency of these conformations is generally lower.

Well-qualified professionals that have comprehensive knowledge of controller behavior and pump curves, are needed to plan such a system. Only an inclusive examination can resolve whether the economical price for such a system makes up for the decline in efficiency and more trouble attaining after-sales service.

6.7.4. The One-Stop-Shop

It looks like that a generally trend is pertaining towards suppliers designing and planning the entire solar powered irrigation system (including irrigation equipment and pump), installing it and posing service contracts for its operation. It is primarily true for larger systems, but there are two exemptions that proves that it is also applicable for the smaller systems as well (e.g., several suppliers in Kenya).

In this case, the One-Stop-Shop business can also work as financier for the entire system. Several solar-powered irrigation companies are concerned with providing the layout and design of the whole system, including preparation of agronomic aspects, and work as a provider of holistic service.

This inclusive view of the solar irrigation system has been accepted widely if one considers 1 other innovative developments illustrated in the following paragraphs.

6.7.5. Performance Monitoring

International solar pump suppliers deliver monitoring and surveillance equipment not only for the operation of the water lifting system, but also for extra functions embracing monitoring and data collection for instance, measurements of water level and/or remote switching of the pump on the basis of standard parameters (water source level, water needs, etc.) This concept has been previously tested with observation wells. The new combined technology makes uses of sensors, that can be incorporated into the production wells and the supervising systems of the solar pumps.

With the help of this information, a groundwater monitoring programme can be framed or designed and the pump capacity can be limited remotely to adapt to a predetermined water level in a way to protect the aquifer. The individual farmer has no control or influence if the Ministry or another institution or authority-e.g., an irrigation association-switches off the pump.

With increased availability of communication technologies, and Internet of things (IoT), pumps can be accessed remotely from anywhere in the world and their overall progress can be measured at any time. Time series of valuable information can be displayed easily.

This allows the supervising of surface and groundwater water levels of essential water resources and their control by dedicated institutions or governments. In addition to this, it also supports reducing water abstraction in the scenario when level of water decline is higher compared with the recharge rate. It is worth noticing that the large irrigation schemes are now making use of these monitoring tools.

They can be introduced further for smallholder farmers having the plot sizes of only around ½ ha. This will provide the suppliers with the opportunity to embark on innovative payment schemes which will make it possible to switch off the whole scheme in case of non-compliance.

6.7.6. IoT Platforms

It is worth noticing that IoT platforms will give SPIS (connected to the Internet, as demonstrated earlier) the prospect to obtain additional services through this platform on the basis of tracked sensors, camera technology and flow meters.

For instance, farmers can get data and information about daily pump usage and weather report, in addition to the recommendations of crop management.

This will no longer be restricted to the larger systems. Solar energy will provide the needed power for collection and transmission of data. For all the new technologies, however, intensive training and education will be needed.

6.8. FUTURE TRENDS IN SOLAR POWERED IRRIGATION SYSTEM

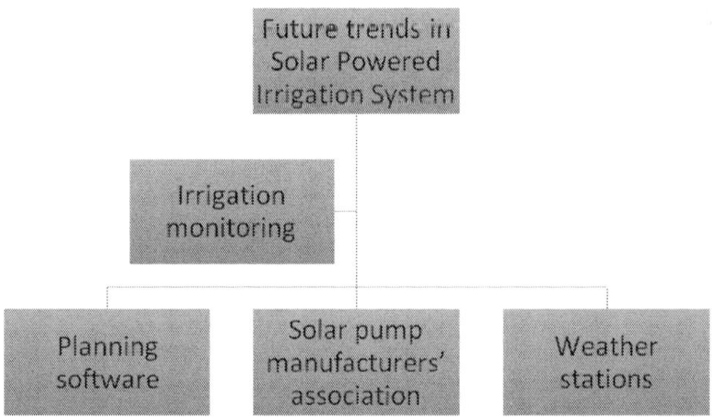

Figure 6.7. Future trends in solar powered irrigation system.

6.8.1. Planning Software

Well-designed, tailor-made and easy to use software is available for solar pumping systems and for several irrigation technologies. However, there is need for assimilation of technology for irrigation and solar pumps and is likely to be developed in the near future. A financing option and sustain efforts to back the execution of these technologies-i.e., without company bias-is instantly needed (Figure 6.7).

6.8.2. Irrigation Monitoring

It is worth noticing that the amount of irrigation water on the field need to be regulated more frequently and regularly. As the amount of water availability gradually decline, this process will play an even more important role.

For this purpose, a differentiation should be made between monitoring water applied, that can be easily governed in pressurized systems with water transpired and flow meters, that can be evaluated through remote sensing technologies measuring biomass production and evapotranspiration. Soon, after a period of time, modern technologies will become increasingly common-for instance, supervising of irrigated fields with thermal imaging cameras and drones.

Thermal and satellite imagery is already used to compute demand of irrigation water (and to measure actual supply) for defined areas (e.g., irrigation schemes, fields, watersheds) and to charge farmers correspondingly. An example of a global database is the FAO Water Productivity Open-access Portal (WaPOR) at 30m, 100m, and 200m resolution for certain countries.

6.8.3. Solar Pump Manufacturers' Association

It can be predicted that solar pump companies will set up their own platform-for instance, a solar pump manufacturers' association. It will allow us to determine standards for the equipment and make the comparison of information and data possible. When successful, smaller companies will likely to join in order to get the benefit from the innovation potential and the data-sharing of such an association.

Many distinctive industries have moved towards this direction and their stakeholders have benefitted from the same. A good example is Regenwassernutzung (fbr) and FachvereinigungBetriebs, the Association for Rainwater Harvesting and Industrial Water Use in Germany, established in 1995.Rainwater harvesting equipment for industrial and housing buildings was not standardized in the 1980s and systems were not compatible. Bringing planners, manufacturers, and users together in an association help in providing much needed fuel to the sector, to control and standardize the quality of equipment, thus develop it further.

6.8.4. Weather Stations

Weather stations are gradually gaining importance for an optimized irrigation schedule and regime. These stations can be developed to a greater extent to become service centers for agricultural crop production as well. It can be possible only if their databases are progressed to provide not only weather data, but data related to soil, such as soil moisture, for the most imperative crops of the region.

Forecasting for the upcoming few days could be easily available, so farmers would be more ware about how much irrigation water to be conveyed for each field and crop. This necessitates the need for a closely linked network of extended weather stations.

6.9. CONCLUSION

In the end, it is concluded that solar irrigation micro-irrigation system is very important, especially for developing countries that relies heavily on agriculture sector. One cannot deny the fact that solar energy micro-irrigation system is cost effective as it does not run on electricity, thus it reduces dependence on electricity in countries where supply of electricity is limited.

However, the initial cost for setup of this system is high, but over a period of time, its overall cost will reduce because of no electricity bill. Therefore, this is considered as the best available option till now if one is looking to implement micro-irrigation technology in their agriculture sector.

In addition, it should be the duty of government to provide loan assistance to poor farmers for helping them to implement solar power irrigation system in their farm land. This will ultimately help the government as they no longer need to provide extra electricity supply to these farmers for irrigation activity.

REFERENCES

1. (2019). Solar-Based Micro-Irrigation System. [ebook] Center for Technology Alternatives for Rural Areas (CTARA) Indian Institute of Technology Bombay. Available at: http://www.ctara.iitb.ac.in/en/system/files/A%20case%20study%20solar%20based%20micro%20irrigation%20system%20.pdf (accessed on 20 August 2020).
2. Deveci, O., Onkol, M., Ozgur, U. H., & Ozturk, Z., (2015). Design and Development of Low-Cost Solar Power Drip Irrigation System Using System Monitoring Language. [online] Research Gate. Available at: https://www.researchgate.net/publication/276426978_Design_and_Development_of_a_Low-Cost_Solar_Powered_Drip_Irrigation_System_Using_Systems_Modelling_Language (accessed on 20 August 2020).
3. Energypedia.info. (n.d.). Sustainable Energy for Pumping and Irrigation-Energypedia.Info. [online] Available at: https://energypedia.info/wiki/Sustainable_Energy_for_Pumping_and_Irrigation (accessed on 20 August 2020).
4. Fao.org. (2020). Solar-Powered Irrigation Systems: A Clean-Energy, Low-Emission Option for Irrigation Development and Modernization. [online] Available at: http://www.fao.org/fileadmin/user_upload/gacsa/docs/GACSA_Practice_brief_SolarIrrigation_v5.pdf (accessed on 20 August 2020).
5. Frometa, R., (2019). 7 Benefits of Having Solar Power Irrigation-Vents Magazine. [online] Vents Magazine. Available at: https://ventsmagazine.com/2019/09/12/7-benefits-of-having-solar-power-irrigation/ (accessed on 20 August 2020).
6. Goyal, M. R., & Ghosal, M. K., (2020). Potential Use of Solar Energy and Emerging Technologies in Micro-Irrigation. [online] Appleacademicpress.com. Available at: http://www.appleacademicpress.com/potential-use-of-solar-energy-and-emerging-technologies-in-micro-irrigation-/9781771883641 (accessed on 20 August 2020).
7. Hartung, H., &Pluschke, L., (2018). The Benefits and Risks of Solar-Powered Irrigation-A Global Overview. [ebook] Available at: http://www.fao.org/3/i9047en/I9047EN.pdf (accessed on 20 August 2020).
8. Irena.org. (2016). Solar Pumping for Irrigation: Improving Livelihoods and Sustainability. [online] Available at: https://www.irena.org/-/media/

Files/IRENA/Agency/Publication/2016/IRENA_Solar_Pumping_for_Irrigation_2016.pdf (accessed on 20 August 2020).

9. Ososanya, D. T., (2015). Design and Implementation of a Solar-Powered Smart Irrigation System. [ebook] Available at: https://www.asee.org/public/conferences/56/papers/13224/download (accessed on 20 August 2020).
10. Rao, K., (2020). Utilization of Solar Power for Operating Micro-Irrigation Systems. [online] ResearchGate. Available at: https://www.researchgate.net/publication/331545897_Utilization_of_Solar_Power_for_Operating_Micro_Irrigation_Systems (accessed on 20 August 2020).
11. Roblin, S., (2016). Solar-Powered Irrigation: A Solution to Water Management in Agriculture?-Renewable Energy Focus. [online] Renewableenergyfocus.com. Available at: http://www.renewableenergyfocus.com/view/44586/solar-powered-irrigation-a-solution-to-water-management-in-agriculture/ (accessed on 20 August 2020).
12. Sumalatha, A., &Eswar, K. G., (2020). Energy Efficient Solar Powered Automatic Irrigation System. [online] Ijemr.net. Available at: http://www.ijemr.net/DOC/EnergyEfficientSolarPoweredAutomaticIrrigationSystem.PDF (accessed on 20 August 2020).
13. Urja Global Limited, (2019). Major Benefits of Solar Water Irrigation. [online] Available at: http://urjaglobal.in/major-benefits-of-solar-water-irrigation.php (accessed on 20 August 2020).
14. Water Dynamics, (2017). The Advantages of Solar Water Pumps for Ground Water-Water Dynamics. [online] Available at: https://waterdynamics.com.au/advantages-solar-water-pumps-ground-water/ (accessed on 20 August 2020).
15. Whitesharkenergy.in. (2018). Solar-Powered Drip Irrigation Technique for Agriculture | Whiteshark-Energy, Efficiency, Environment. [online] Available at: https://whitesharkenergy.in/blog/solar-powered-drip-irrigation-technique-for-agriculture/ (accessed on 20 August 2020).

Chapter 7

Sustainable Micro-Irrigation

CONTENTS

The chapter of sustainable micro-irrigation explains the basic significance of the sustainable micro-irrigation. This chapter also explains the pressurized irrigation technology that are used in irrigating the fields. This chapter also addresses the factors that impacts the adoption of pressurized irrigation technology among olive farmers.This chapter also provide highlights on the affordable or cost-effective systems for micro-irrigation for small farmers in emerging nations. This chapter also emphasize designs of low-cost micro-irrigation systems, uniformity of the emitter flow, and global application of these sorts of low-cost micro-irrigation systems.

This chapter also consisting of low-cost pipes distribution system, system layout of the low-cost pipes distribution system, components of the low-cost pipes distribution system, as well as the design criteria and considerations. This chapter also explains the significance of subsurface drip irrigation, advantages of the subsurface drip irrigation, components of the subsurface drip irrigation units, disadvantages of the subsurface drip irrigation, and field installation of the subsurface drip irrigation units.

7.1. INTRODUCTION

The field of agriculture is one of the major sources of livelihood for farmers all across the globe. Farmers or agriculture contributes to the nation's gross domestic product (GDP) by near about 30%, which is kind of huge contribution.

According to the reports of CWC (Central Warehousing Corporation) which states that the utilization of the water for the purpose of irrigation, accounted for near about seventy- five per cent, followed by the domestic use of near about ten per cent, industries by five per cent, and other activities claimed near about ten per cent, which is consisting of evapotranspiration losses, environment losses, environment, and navigation requirements.

Near about 80% of the water which is being utilized with the help of the irrigation sector is decreased to near about 60%, and the reason behind this is that the competing water demands from the other sectors.

The efficiency of the irrigation fluctuates from 55% to 90%. The efficiency of the irrigation rely on the various kinds of techniques that are being adopted for the process of irrigation, for example, flood irrigation give near about 65% of the water use efficiency (WUE), furrow irrigation gives near about 80% WUE, and micro-irrigation gives near about 85% to 90% WUE (out of which, sprinkler give 80%, and drip gives 90% WUE).

Initiatives shall be taken for enhancing the overall water use effectiveness in the field of agricultural sector for surface water system and groundwater system from that are ranging from 30% to 65% and 65% to 80%.

Achieving high WUE is one of the most important steps in the direction towards the sustainable water development and management. Precision irrigation is one of the techniques in order to achieve the sustainable agricultural management.

There are certain limitations that are required to be addressed for consumptive water use in agriculture, for example, decrement in the water losses in the conveyance and distribution system with the help of the periodic maintenance, application of water to the roots of the plants directly with the help of sprinkler, drip, and the methods of landscape irrigation, deficiency of awareness between the farmers with respect to the new, modern, and innovative technologies, insufficient networking between the farmers, government, and the private sector providing precision irrigation methodologies, deficiency of proper profits, and subsidies for modern irrigation technology, adaptation between the farmers from several state governments, and the risk of the crop failure between the farmers with the help of the adopting new modern irrigation technology (Figure 7.1).

Figure 7.1. The basic significance of sustainable micro-irrigation.

Source: Image by Pixabay.

The reformation of the sector of the irrigation necessitates the enhanced policy measures in order to create synergy between the decisive water governance institutions, for example, the Central Water Commission, State Irrigation Department, National Rainfed Area Authority, and Central Groundwater Board in order to make sure the sufficient amount of water is available for the poor as well as marginal farmers for agricultural use with the help of decreasing the legal pluralism problems and water conflicts regarding the water equity that are based on the gender equality.

The best scientific practices should be adopted for enhancing the WUE in the field of agriculture such as proper management of the cultural command area size based on water requirement for the crop, time linkage between the water supplies for irrigation, and plant water requirement through the course of the Khor period, the selection of the crop is based on the condition of the climate as well as the type of the soil.

In addition to this, appropriate provision in order to avoid the extra amount of the water wastage on fields, minimize the evaporation of the water losses from the irrigated fields with the help of the mulches, adaptation of the precision irrigation methodologies, for example furrow irrigation, pressurized irrigation (drip or sprinkler irrigation) under the subsidy to the farmers, usage of the hybrid seeds in drought-pone regions for drought-tolerance crops, "more crop per drop" strategy should be adopted.

More specifically, participatory irrigation management promoted involving, non-profitable government organizations (NGO), the local community or local groups of the farmers, as well as the local government administrative body, consumptive usage of the surface and groundwater.

Adopting the all above-mentioned techniques and methodologies, an objective of a sustainable agricultural water management will be accomplished, in this way, the augmenting the overall WUE by near about 75% to 80% in the agricultural sector.

7.2. PRESSURIZED IRRIGATION TECHNOLOGY

With respect to the pressurized irrigation systems, the water is being pressurized and precisely applied to the plants under pressure with the help of system of pipes. Pressurized irrigation systems, as opposed to the surface irrigation systems, are more efficient in application of the irrigation water to the crops.

They offer enhanced farm distribution, enhanced control over timing, decrement in the wastage of the land in laying field distribution network, decreased demand for labor and better use of the limited resources of water.

There are several numbers of variations of pressurized irrigation systems but, out of them, there are two major pressurized irrigation systems. The first one is drip irrigation systems, and the second is sprinkler systems. Out of them, there are several numbers of variations according to the type of the field, the crop and the type of the water delivery fittings that are required, but the components of the basic system remain the same.

They are consisting of the control station where the pressure is applied, the mains and the sub mains, that are the pipes of various lengths and dimensions, the manifolds or feeder pipelines, and the laterals or the irrigating pipelines that are equipped with the emitters (more particularly, in the case of drip irrigation) or the other fixtures that provide assistance in delivering the water to the plants or spray booms for the forage and grains and the center-pivot systems (specifically, in the case of sprinkler irrigation systems) (Figure 7.2).

Figure 7.2. Pressure irrigation technology can be used for both drip as well as sprinkler irrigation system.

Source: Image by Pixabay.

Fertilizer injectors can be used into the pressurized irrigation systems in order to perform the process of fertigation. Pressurized irrigation systems have the capabilities in order to prevent the loss of water which is related to the surface irrigation elevating the open irrigation application efficiency ranging from 45% to 60% to pressurized irrigation with the efficiency of range starts from 75% to 95%.

On the other hand, the open canals systems necessitate high labor in order to maintain the pressure in the pressurized systems. Pressurized irrigation systems require from 1.10% to 25% of the man hours canal systems necessitates.

Driven by the requirements in order to decrease the input of labor in agriculture and the love of high technology, pressurized irrigation systems are very expensive in nature and out of reach of small-holder farmers in the emerging nations.

The quality of water and the energy are the most important aspects that plays a crucial role with respect to the sustainability of the pressurized irrigation systems. Water with high dissolved minerals leads to the continuous hindering of emitters. Regular or frequent maintenance is required in order unblock the delivery fittings and in order to maintain the pumps and fertigation units.

The pressurized irrigation systems are ranges from sophisticated computer-controlled setups with the high start-up costs to moderate-cost system that is consisting of bubblers, mini-sprinklers for orchards and micro-sprinklers.

Their utilization has been greatly restricted to the large farmers or in commercial farming.

Pressurized irrigation systems for small-holders have recently been popularized. Some examples of low-cost systems are drip irrigation with the drip emitters, drip tape and porous-wall pipes, along with the hand-moved hose basin system for trees, pipe distribution irrigation systems (PDS) and hose-move sprinklers.

7.3. FACTORS AFFECTING ADOPTION OF PRESSURIZED IRRIGATION TECHNOLOGY AMONG OLIVE FARMERS

There are persuasive discussions which are in favor of planting and growing olive trees and creating the related products all across the globe. The main reason of this popularity does not only route in the agronomic and economic reasons. Never the less, the environment as well as the human health are some of the most important causes that are required to be well thought out.

This tree is significantly tolerant to drought, and it is having the potential to be grown in soils which suffer from poor quality and are shallow.

These aspects or characteristics put this species among the most interesting ones or cultivation in both arid regions as well as semi-arid regions (Fernandez and Moreno, 1999).

Talking about Iran, and more particularly in Roudbar County, agriculture plays an important role with respect to the socio-economic development. It is quite obvious that the enhancement in the agriculture is highly dependent on the water. The country of Roudbar enjoys from a Mediterranean climate.

In these kinds of climate, agriculture is the major and the most essential consumer of water and a significant percent of available water resources as well above 80% is allocated for the process of irrigation (Laraus, 2004). Celano et al. (1999) believe that olive trees are more respondent with respect to the irrigation management plans.

Because of the fact there are increment in the demand, the limited resources water table variation in space and time, as well as soil contamination, it is needed to regulate the available water resources at farm level optimally (Kumar and Singh, 2003).

A very necessary element is effective operation and the management of the irrigation schemes is effective water management. In simple words, with the help of the irrigation water management is identifying at which time to irrigate, the amount of water to utilize at each and every irrigation stage and through the course of the plant, also operating and maintaining the irrigation system.

The main reason of the management system is to regulate the production system for profit without doing any sort of harm to the environment so that the management is also in agreement with water availability plans.

One of the most important concern in the management programs is to determine when and how much water is required in order to apply or irrigation. With respect to this, the method of irrigation as well as the other field characteristics must be well thought out very carefully (Holzapfel et al., 2009).

As there has been a significant decrement in the production of the olive in the region over the time period of last ten years and conventional methods for irrigation in olive groves, it is very important in order to study the pressurized irrigation technology adoption among the farmers of olive in Roudbar County.

The significance of this matter is not just limited to this field of study, and several numbers of various studies have been accepted in this regard in Iran and other places as well, all across the glove in the time period of last few decades.

According to the study done by Movahedi et al. (2017), which was focused on determining the factors that are impacting the adoption of technology of pressurized irrigation by the farmers in in Asadabad County found that the variables are consisting of "perceived usefulness," "perceived ease of use" and "attitude to use" had a considerable and positive effect on the decision in order to accept the use of the pressurized irrigation systems.

In addition to this, according to the Afrakhteh et al. (2014), in their studies, that was related to the analysis of the factors that are impacting the adoption and the use of the sprinkler irrigation by the farmers in the Famenin County concluded that the adoption of the sprinkler irrigation systems to be impacted with the help of the environmental factors, for example, the region under cultivation, access to the water, the quality of the water, and non-environmental factors, for example, the workforce number in the family, employment diversity and participation in extension education and courses on agricultural water management.

According to the study results of the Adeoti (2009) that shows the factors that are impacting the technology of the irrigation adoption and its influence on the household poverty in Ghana shows that the availability of the labor and the increment in the number of extension visits per year are the factors that elevated the possibility of the adoption.The prime objective of this descriptive correlational study was to find out the influencing factors on adopting the technologies from the olive farmers in the Roudbar County, Northern Iran, with the help of partial least squares (PLS) technique, using Unified Theory of Acceptance and Use of Technology (UTAUT).The technology acceptance models (TAM) try to describe the level of acceptance of the application of the technologies. These theories evaluate whether the user will be able to accept the new technologies and the ability of the user in order to deal with them.

The TAMs provide assistance to the management executives as well as decision-makers in order to evaluate the effectiveness of the introduction of the technology to the organization and motivate the user in order to accept the systems. Unified Theory of Acceptance and the Use of Technology (UTAUT) is a user acceptance model which is developed by the Venkatesh et al. (2003).Unified Theory of Acceptance and the Use of Technology is one of the new models in the area of TAM, and its main cause is to describe the intention of the user in order to use the technology and in order to inspect the usage behavior (Taiwo and Downe, 2013). The Unified Theory of Acceptance and the Use of Technology altogether develops across the eight models.

The eight models are consisting of, as mentioned below:

1. The Technology Acceptance Model (TAM) (Davis, 1989);
2. The Theory of Reasoned Action (TRA) (Fishbein and Ajzen, 1975);
3. The Theory of Planned Behavior (TPB) (Ajzen, 1991);
4. The Combined TAM and TPB (Taylor and Todd, 1995);
5. The Innovation Diffusion Theory (IDT) (Moore and Benbasat, 1991; Rogers, 2003);
6. The Social Cognitive Theory (SCT) (Bandura 1989; Compeau and Higgins, 1995);
7. The Motivational Model (MM) (Davis and Warshaw, 1992); and
8. The Model of PC Utilization (MPCU) (Thompson et al., 1991).

The Unified Theory of Acceptance and the Use of Technology is consisting of four core determinants of technology use behavior and up to four moderators of key relationships.

7.4. AFFORDABLE MICRO-IRRIGATION SYSTEM FOR SMALL FARMERS IN DEVELOPING COUNTRIES

In the regions where the availability of water is not so common, drip irrigation offers the most effective way in order to save the water for irrigation but its cost of near about a thousand US dollars for an acre is prohibitive for most of the small farmers in the emerging nations.

The cost was decreased by up to 90% with the help of:

1. Making dripper lines moveable, in such a way that each and every line reaches ten rows in spite of one;
2. Substituting 25 cm with simple 0.70 mm holes punched with the help of a heated needle; and
3. With the help of $3.00 off-the-shelf 20 liters containers with cloth filters in place of expensive filter systems.

This decrement in the cost of a half-acre system to fifty dollars. The low-cost system was field-tested in the hill stations of Nepal, and in mulberry cultivation in the regions of Andhra Pradesh, India. Uniformity of flow from emitters was near about 73% to 84%.

Small farmers reported that the low-cost drip irrigation system reduces the requirement of labor for the irrigation system by nearly 50%, two folds the area irrigated with the help of same amount of water. The low-cost drip system is likely to be extensively adopted with the help of the farmers in semi-arid and hilly areas.

In all across the globe, several numbers of irrigated regions are running increasingly short of water (Bucks, 1995). With the increment in the competition among the farmers who require to irrigate their crops and the rapidly growing cities of the world, water saving techniques of irrigation are at a premium (Postel, 1992).

Drip irrigation is one of the most effective type of irrigation, which is available at present interval of time, now irrigates a little under two million hectares. This is less than 1% of the global irrigated acreage. The major blockade to its extensive adoption is its high capital cost (Postel, 1996), which is near about a thousand US dollars per acre for row crops such as vegetables.

This makes the installation of the drip systems prohibitive for the great majority of the farmers in the emerging nations who cultivate less than five acres.

Through the point of view of the economists, at the present interval of time, the technologies that are being used in the irrigation are not "divisible," that is it cannot be separated into very small functional units. A typical small farmer in India cultivates five plots that are ranging in size from one-fifth of an acre to a half-acre.

Biological technologies such as Green Revolution seeds and fertilizers, are divisible. And the reason behind this is that they can be divide into small affordable packets that fit tiny plots. But the drip irrigation systems, such as other current irrigation technologies, are very difficult to customize in order to fulfill the requirements of the small plots of poor farmers, and they are very expensive in nature to be affordable.

The development of a reliable low-cost drip system that fulfils the requirements of the small farmers in the emerging nations has long been recognized as an important necessity (Hillel, 1985; Saksena, 1995; Nir, 1995). For instance, Sivanappan, and his colleagues tested the usage of the holes in spite of emitters in the India in the time period of 1970s (Sivanappan and Padmakumari, 1980).

In order to fulfill the requirement for a low-cost system, International Development Enterprises (IDE), which is a non-profit organization having the specialties in small scale irrigation, has created and field tested a simplified drip irrigation system that is divisible and costs near about two hundred and fifty US dollars a hectare (near about $50 for a one-half acre unit).

7.5. DESIGN OF LOW-COST MICRO-IRRIGATION SYSTEM

The prime contributors to cost for a standard drip irrigation system are mentioned-below:

The system of drip lines delivering water to each and every row of plants;

Hundreds of 25-cent plastic emitters providing the water at each and every drip point; and

A filter system at the water intake point.

With respect to the small farmers, the IDE have decreased the capital investment in drip lines with the help of making them mobile. One movable drip line serves ten rows of plants in spite of one. Emitters are substituted with the help of the holes in the plastic emitter lines that are made by 0.7 mm heated punch, which is exactly the same size as an ordinary safety pin.

Small farmers can then use a safety pin in order to manually clear any sorts of holes that become plugged. In order to prevent the spraying at each and every hole, a simple baffle is fabricated from a six cm length of drip line splits in horizontal direction and snapped over the hole.

The costly filtration system of the standard drip systems is substituted with the help of off-the-shelf, three-dollar twenty liters plastic tank situated at a height of two meters above the ground level, with a nylon cloth which act as a filter at the water entry point.

These various kinds of modifications decrease the cost of a system from $ 2,500 to $ 250 per hectare. A proof-of-concept prototype system was fabricated in Nepal functioned satisfactorily and was followed by the trails carried in the fields of farmers.

7.5.1. Uniformity of Emitter Flow

On the other hand, several numbers of various field studies have verified the superior performance of drip irrigation systems in economizing the use of the water (Chandia et at., 1995), assessments of its potential in order to apply water uniformly have produced mixed results (Hansen et al., 1995).

For instance, out of nineteen, eleven of the drip systems assessed in the Florida had emission uniformity less than 90%, when they are compared with only twelve out of fifty-seven micro sprinkler systems (Yurgalevitch et al., 1995).

The Bureau of Indian Standards (BIA) has set a standard of no more than 5% variation for class A emitters. Manufactures in India, such as, those in the other nations, have found great difficulty in order to fulfill or follow this standard (Chauhan, 1995).

In the recent interval of time, the American Society of Agricultural Engineers (ASAE) has specified standards as "excellent" for coefficient of variation for emitters of less than 5%, and "average" for variation 11% to 15%.

Tests of uniformity in the IDE low-cost drip systems found a variation from 12% to 29% from hole to hole at heads fluctuating from two to four meters.

Several numbers of various studies of standard drip systems have found that the clogging of drippers and the fluctuations in line pressure make considerable contributions to the emitter variance.

For instance, the assessment of the standard drip systems that are installed in the Indian State of Maharashtra found emission uniformity of less than 50% in the installations in the areas that are steep or hilly, more specifically, if the systems used an unsuitable pump set and contour mapping was not used (Holsambre, 1995).

The IDE low cost system employs simple color-coded pressure reducers. These color-coded reducers can be installed with the help of the farmer in order to balance the pressure for lower terraces in hilly regions.

The holes of the drippers are designed in such a way that they all are of same diameter just like to the standard safety pin and farmers are taught to frequently check drip lines and clear the holes that may become plugged. The simplicity of clearing the plugged holes make it possible in order to decrease the investment in the filter systems.

It could be expected that the ease of clearing the plugged holes that are integrated with the help of color-coded push-on pressure controllers will give benefits for longer period of time in emitter discharge uniformity, but also, this can only be identified with the help of more extended field tests.

7.5.2. Global Application

The modern surface micro-irrigation developed from the Israel in the time period of 1960s. And, in the present interval of time, the nation irrigates near about two million hectares. Decreasing its capital cost from two thousand and five hundred US dollars to two hundred and fifty dollars a hectare will likely at least two folds its global adoption, and the adoption curve for low cost trickle system will favor the small farmers.

Moreover, to mulberry and vegetable crops, the low-cost systems will me more attractive for the small farmers who are cultivating the herbs, horticultural crops, cotton, along with the sugar cane. More particularly, the application of low-cost systems will be in that regions where the irrigation water has high value, for example, semi-arid regions in India and Sub-Saharan Africa.

It is quite suitable for the small farmers who lives in hilly regions for example, Himalayan hill areas, as well as semi-arid regions such as Sub-Saharan Africa. More particularly, the low-cost system will be helpful for canal systems producing unsatisfactory water in order to cover irrigable land in the canal distribution area (Sawleshwarker, 1995).

7.6. THE ADOPTION OF WATER CONSERVATION APPLICATION TECHNIQUES

There are several numbers of farm application irrigation technologies that could provide assistance to the conservation of the water with respect to both, surface systems as well as piped systems, have been developed. Generally speaking, these kinds of techniques or methodologies are adopted for the cultivation of the high value crops.

All across the globe, near about twenty million hectares are now irrigated with the help of sprinkler irrigation, Micro-irrigation as a technology has matured into a reliable water and fertigation management system for the cultivation of crop over the time period of last ten years and its application continues to increase with quick pace.

A total of near about 3.2 million hectares are being irrigated with the help of techniques of micro-irrigation all across the globe which shows that 1% of the total areas that have been irrigated in the world. A development of over six folds over the time period of last two decades has been accomplished, which is quite a success.

Micro-irrigation is a very significant tool in the drive towards food security. Micro-irrigation systems have great capabilities in order to produce considerable opportunities for small farmers and small-holder agriculture.

Through the course of the time period of 1980s, micro-irrigation started penetrating the emerging nations, more particularly in India. After few years later, fearing a shortage of water and shortage of food, the Chinese Government declared the technology of the drip irrigation to be among the top priorities for the development in the time period of last ten years (Figure 7.3).

Figure 7.3. Sprinkler irrigation system: One of the most effective water conserving irrigation techniques.

Source: Image by Flickr.

Never the less, However, Vidal et al. (2001) have shown that micro-irrigation "is not a miracle technology: excellent as well as poor results were gained with the help of these technologies, and their adoption rally relies on the potential of the farmers in order to finance and operate them, as well as is relies on the type of the crop production."

As per the database of the International Commission on Irrigation and Drainage (ICID) on modern irrigation techniques, it gives the impression that the use of the pressurized water application techniques is extensively different between nations.

The percent of the modern technologies reduces with the total area irrigated in emerging nations from a low of 21% in the United States of America and 16% in Italy to 90% or more than 90% in the Northern European nations (such as France and Germany).Sprinkler has a much greater utilization when it is compared with the drip systems with the exception of Cyprus, Israel, and Jordan (Table 7.1).

Table 7.1. Sprinkler and Drip Irrigation in Selected Countries (The data refer to different years between 1993 and 2000)

Country	Sprinkler irrigation (million ha)	Drip irrigation (million ha)	Total area irrigated by modern methods (million ha)	As percentage of total irrigated area
China	1.20	0.27	1.47	2.8
Cyprus	0.002	0.025	0.027	49
France	1.40	0.05	1.45	90
Germany	0.53	0.002	0.532	100
Israel	0.07	0.16	0.23	100
India	0.66	0.26	0.92	1.6
Italy	0.35	0.08	0.43	16
Jordan	0.005	0.038	0.043	62
South Africa	0.26	0.22	0.48	37
USA	3.38	1.05	4.43	21

Source: Table from fao.org

In the region of South Asia and Africa, the low-cost bucket and drip sets are becoming more and more common with the farmers. In the regions where shallow ground water is available in sufficient amount, thousands of poor farmers from Bangladesh have used low-cost treadle pumps in order to supply the water for crops for their own food security and additional income.

The enhanced techniques and methodologies for surface irrigation, for example, precise land leveling, gated pipes or can legation have also been developed in order to enhance the typical poorly distributed application of the water under traditional methods with the help of furrow, border or basin irrigation.

Experiences shows that these kinds of techniques are very cultured to be adopted by the most of the smallholders in the emerging nations. Adaptive research in the Middle East as well as North America has shown the challenges in introducing the surface water conservation techniques.

Several numbers of attempts have been tried in order to introduce the gated pipes (for instance, in Morocco) have been failed for several numbers of reasons. Out of which, the main reason is the poor maintenance of the quaternary earth canals. Significant attempts for training of farmers in the conservation of water in surface systems are required.Farmers have adopted the pressurized system much faster when it compared with the surface application techniques. For instance, near about 65% of the farmers in the Jordan valley have changed their irrigation system from surface irrigation to the drip irrigation, over the time period of a decade. Farmers build on-farm storage reservoirs in order to give the flexibility that is needed for the drip irrigation.Never the less, there are no statistical data available on it, it is likely that most of the regions irrigated with the help of the techniques such as conservation of water, are served from wells or piped systems. In the present interval of time, there is still a huge potential for the development of techniques of conservation of water in the regions that are irrigated from groundwater.Application of these kinds of techniques will continue to develop with the market for high value crops. Never the less, the adoption of these kinds of techniques are restricted to the cultivation of specific kinds of crops and the reason behind this is the technical and financial consideration.

7.7. LOW-COST PIPES DISTRIBUTION SYSTEM

Pump irrigation covers most of the irrigated lands in the emerging nations of the arid as well as semi-arid areas. The water for the irrigation taken from the aquifer or the surface run off ponds, lakes, rivers, and dams, is pumped to the fields with the help of the traditional (earth) ditches or lined canals, as an outcome, huge amount of losses from seepage and evaporation, deep percolation and canal leakage.

According to the several numbers of studies, from various nations shows an average of 33% water losses through the course of the conveyance with the help of a hundred m conventional channel. The methodologies for field irrigation are the conventional surface gravity-furrow, basin, border, and so on, with the help of field application effectiveness of 60% to 70%, that is, additional water losses of near about 20% to 27% of the total.

Then the overall irrigation efficiency ranges from near about 40% to 47%. The solution to the issue is the closed pipes enhanced irrigation techniques. The huge gap between the water wasteful open surface irrigation practices and the highly effective enhanced irrigation techniques can be removed with the help of the execution of the Pipe Distribution Irrigation System (PDS).

This method of irrigation has been widely used in Cyprus in the time period of early 1960s and in Yemen (particularly in Tihama) in the time period of early 1980s with big success. It is actually the substitute of the open channel with a suitable designed closed piping network in order to pass and spread the irrigation water up to the field plots without any kind of losses (Figure 7.4).

Figure 7.4. An illustration of low-cost pipe distribution system.

Source: Image by Flickr.

It is a quite simple technology with the minimum cost, which in one night may increase the overall efficiency of the irrigation from 40% to near about 77%. The PDS is a closed pipe surface irrigation tactic and it is categorized as a low medium pressure system with solid permanent installation.

7.7.1. System Layout and Components

The basic layout of the piped distribution irrigation system is consisting of a simple control head, a pipe distribution network and the hydrants.

The control head is consisting of the essential regulating valves (shut off, check valve, air valve) which are placed on a piece of galvanized steel threaded pipe, sixty meter above the surface of the ground, with tee outlet for taps and pressure gauge. This arrangement, at a later stage, can easily be transformed into a more sophisticated control unit, suitable for micro-irrigation systems.

The main and sub-main pipelines (which is known as distribution network) can be of rigid Poly-vinyl chloride (PVC), 90–160 nm DN, 4–6 Bars PN, placed under the ground. On hilly regions, other types of pipes are used on the surface of the ground, for example, flexible back poly-ethylene (HDPE), the Quick Coupler light steel or the galvanized steel threaded pipes. The latter is used only up to the three in size, because of its expensive nature.

The hydrants are rising on surface that are equipped with a shut off valve (which is known as valve).

They are capable in order to deliver part of the systems flow or the whole of its to the manifold open ditches. At the later stage portable lightweight pipes (quick couplers aluminum, light steel, lay-flat hose, black polyethylene, and so on) can be attached to the hydrants, substituting the manifold open ditches, for the final delivery.

From the hydrants the irrigation water is released directly to the manifold open earth channels for the diversion to the furrows, the basis or the strip borders.

7.7.2. Design Criteria and Considerations

The pipes distribution system is consisting of both the feature of the open surface methods as well as the pressurized closed pipe technology. The design standards and the criteria are too many when they are compared with the simplicity of the installation.

The structure of the area (such as shape, slope, and so on), the type of the soil, the size of flow, and the method of water delivery to the crop (such as basin, furrow, border or other) should be strictly evaluated.

The take-off hydrants must be positioned at the highest points of the fields plots and at the right distances in order to allow the effective practice of the gravity irrigation techniques with the help of the manifold ditches.

"The most essential measures or the standards that are needed to look out through the course of the design is the likelihood for future expansion of the network for the adoption of any other low-medium enhanced irrigation system, for example sprinkling, drip, spitters, and so on, with the minimum expenditures.

Then the careful design of a flexible skeleton-piping network, that are appropriate in order to fulfill all the techniques of irrigation and water delivery techniques is of the prime importance."

Table 7.2. Flow Versus Pipe Diameter

V = 1.7 m/s²	Galvanized steel threaded, light series			Rigid PVC, 6.0 bars			HDPE, 6.0 bars	
DN mm	2 in	2.5 in	3 in	90	110	160	90	110
Di mm	54	69	82	84.4	103.2	150.2	79	97
Flow m ³ / h	14	23	32	34	51	108	30	45

Source: fao.org

1. Cost

This system is low-medium pressure strengthened with solid permanent installation. The initial cost per unit of area is lesser when it is compared with the other closed pipes pressurized irrigation systems. The average cost per hectare is near about eight hundred and fifty US dollars.This system is a low-medium pressure with solid permanent installation. The initial cost per unit of area is lower as compared with the other closed pipes pressurized irrigation systems. The average cost per ha is around 850 US dollars.In addition to this, the consumption of the fuel is lower than in any other enhanced irrigation system. Only the labor expenses are comparatively high. This methodology or techniques is categorized as low-cost irrigation technology.

2. Advantages

1. Low initial capital investment;
2. Availability of equipment;
3. Easy to operate and maintain;
4. High adaptability by the farmers;
5. Suitable for a wide range of crops; and
6. Appropriate for all sizes of holdings and plot irregular shapes.

3. Disadvantages

1. Necessitates skilled irrigator;
2. Low field application (or delivery) efficiency;
3. In unleveled fields and sandy soils is not easily adapted;
4. Not applicable for small flows; and
5. Best suited for medium-heavy soils.

In spite of the advantages as well as disadvantages that are mentioned above, the main aspect of the system is that this is the first step to be taken in order make it much easier for the farmers to shift from conventional irrigation processes to the more enhanced irrigation practices, that are safe and much more cost effective as well.

4. The System Layout and Dynamic Head

There is transportation pipeline, which is near about 100 m long from the pump to the farm gate. This transportation pipeline is made up with the help of the rigid PVC 90 mm, 6.0 Bars. The on-farm piping distribution network, of the same type and same size, is placed with the borderlines of the field plots (Table 7.3).

Table 7.3. List of Equipment Needed for the Pipe Distribution System Installation (Bill of quantities). Area: 3.3 ha, System flow: 27 m^3 /h, Dynamic head: 0.9 Bars. 0.9 Bars.

Description	**Quantity**	**Unit price US$**	**Total price US$**
System distribution network			
90 mm rigid PVC pipe, 6.0 bars	660 m	2.50	1650.00
2.5 in x 90 mm PP male adaptor	1 pc	10.00	10.00
90 mm PP elbow	1 pc	15.00	15.00
90 mm PP tee	5 pcs	22.00	110.00
90 mm x 2.5 in PP female tee	6 pcs	19.00	114.00
90 mm x 2.5 in PP female elbow	6 pcs	13.50	81.00
2.5 in threaded riser pipe 60 cm	12 pcs	5.00	60.00
2.5 in brass gate valve	12 pcs	14.00	168.00
2.5 in elbows threaded male	12 pcs	3.00	36.00
Trench excavation and backfilling	660 m	1.00	660.00
Sub-total			**2904.00**
Head control			
2.5 in brass check valve	1 pc	16.00	16.00
2.5 in brass gate valve	1 pc	14.00	14.00

2.5 in tee threaded female	2 pcs	3.50	7.00
2.5 in nipple	2 pcs	1.00	2.00
2.5 in pipe threaded 60 cm	2 pcs	5.00	10.00
2.5 in elbows threaded female	1 pc	3.00	3.00
1 in single air valve	1 pc	12.00	12.00
Pressure gauge w/adaptor base 2.5 in	1 pc	14.00	14.00
Sub-total			**78.00**
TOTAL COST			**2982.00**

Source: Table from fao.org

7.8. SUBSURFACE DRIP IRRIGATION

About the spread of Subsurface Drip Irrigation (SDI) there is a general agreement. As the areas under the SDI are counted as drip irrigation in most surveys therefore it is difficult to obtain this trend.

Further, SDI comprises only about 27% of the land area which is basically devoted to drip irrigation and SDI this basically inferred in the US by the means of USDA Farm and Ranch Irrigation Survey. This survey basically provides a better understanding towards this rising trend.

Further it has been observed that as the farmers substitute their conventional drip irrigation methods by the SDI systems therefore this percentage is continuously rising over drip irrigation. Moreover, a potential increase could be easily observed due to the SDI system in the future if this framework continues at the same pace. And further this potential increase could be remarked tremendous.

The SDI with rudimentary laterals and the low emitter pressures which highlights even more basically in the underdeveloped countries there are many references regarding these so-called low-cost SDI. Further this SDI system stands out well and enhances the potential increase leading towards tremendous results. By the saving on the water and energy likewise this has been enhanced.

7.8.1. Advantages of Subsurface Drip Irrigation

The drip irrigation methods are considered as the vast methods as they provide increased crop yield with less labor and reduces the labor cost for the farmers. And many more experimental evidences of SDI advantages over the irrigation methods show a vast aspect. Further, Lamm, and Payero compiled the advantages and thee drawbacks of this method and some of them are basically mentioned below to get a better clarity of the system.

As the surface runoff, soil evaporation and deep percolation are greatly reduced by the system or nearly eliminated therefore it has been observed that the efficiency of water use is comparatively high. Further, as the movement of fertilizers and other chemical compounds by the means of deep percolation is reduced therefore, it leads to the risk of the aquifer contamination is decreased which again shows a tremendous result over all.

Further by using such an advanced system pathogen drift, human, and animal contact with such waters by the use of degraded water, surface waste water application.Moreover, as the fertilizers and pesticides can be applied with accuracy leads to the efficiency in water application is improved. The reducing unnecessary irrigation water losses due to widely spaced crops, a smaller fraction of the soil volume can be wetted. In drier regions as the reductions in weed germination and weed growth.Having reduced manual exertion and injuries due to the hand laborers benefits from drier soils. Less than in drip irrigation mainly due to the operating pressures.Further improvement in the double cropping opportunities which was observed. Since the system need not be removed at harvesting or reinstalled prior to planting the second crop therefore it led to the crop timing could be enhanced. The potential for vandalism is also reduced and also the lateral and sub mains can further experience less amount of damage on the other hand (Figure 7.5).

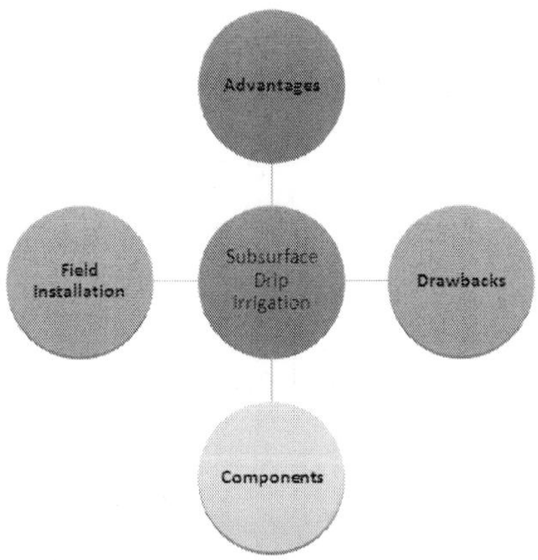

Figure 7.5. Subsurface drip irrigation.

7.8.2. Drawbacks of Subsurface Drip Irrigation

It is more difficult to evaluate system operation and the water applications uniformity and also the water applications may be largely unseen. Further there are other disadvantages are caused which could led to hamper the productivity output and these may be the under irrigation, less crop yield quality reductions and over irrigation and all of these disadvantages could be caused by the means of the system mismanagement. Further problems such deep percolation problems and poor soil aeration could be caused.

Further the enhancing surface and undesirable wet spots in the field could be caused if the emitter discharge exceeds soil infiltration, a soil overpressure which will develop around the emitter outlet. There are explicitly many requirements which should be focused upon and these are namely the timely and the consistent maintenance and the repairs.

Further, particularly for the deeper SDI systems the leaks which are mainly caused by the rodents are difficult to locate and repair. To limit or prevent further intrusion the design procedures must employ safeguards. In addition to this, also for the root intrusion the drip line must be monitored.

The flows are reduced or eliminating by the means of the roots from some perennial crops which may pinch drip lines. In order to remove accumulation of silt and other precipitates which basically occur in the laterals could be periodically done by the means of the drip lines which need to flush them out.

Other than the alternative irrigation systems there is an explicit need of likewise operation and management which further requires more consistent overnight. The air relief and vacuum breaker devices must be present and be operating correctly therefore it leads to the possibility of soil ingestion at system shutdown. At some depths, causing adverse effects on the flows the compression of laterals by the soil overburden can occur in some soils.

7.8.3. Components of Subsurface Drip Irrigation Units

It is a looped network as the typical SDI unit, contrarily to drip irrigation units. All the laterals are connected that it is connected by a sub main of feeding pipe, and a flushing pipe.

The flow variability within the unit is also improved the looped network is enhanced by the means of hydraulic variability. The water is mainly supplied directly into the crops root areas the pipes and laterals are buried at a certain depth which is generally from 7 to 40 cm.

The water suspended elements and possible sediments by the flushing pipe eliminates. The air at the beginning of the irrigation and let it pass when the irrigation basically stops as the two relief valves may be placed at the head and the tails of the feeding and the flushing pipes.

Further into the laterals, the soil particles are basically introduced. Developed for that purpose alternatively some flushing valves for the lateral ends.

The emitters in the generally impregnated with a weed killer to reduce the root intrusion at the emitter outlet as the emitter in the SDI system is similar to the drip irrigation. With the specific geometries to reduce the risk of the root intrusion the countries where herbicides are forbidden, commercial emitter models are basically developed.

Further to avoid the entrance and the deposition of the soil particles likewise there are some of the models along with the models with elements such as diaphragm and geometries.

7.9. CONCLUSION

Micro-irrigation is an effective technique in order to provide water in agriculture. The micro-irrigation system permits a continuous application of water with the help of drippers at particular locations on the lateral lines it permits satisfactory or advantageous settings for soil moisture in the root zone and the optimal development of the plant.

A well-designed system of micro-irrigation can elevate the production of crops. And the reason behind this is the effective use of water, enhanced microclimate around the root zone, pest control, weed control, as well as agronomic and economic advantages. Hydro-Calc, with respect to its properties that shows a dependable tool in designing a system for micro-irrigation.

REFERENCES

1. Food and Agriculture Organization, (n.d.). The Adoption of Water Conservation Application Techniques. [online] Fao.org. Available at: http://www.fao.org/3/y4854e/y4854e06.htm (accessed on 20 August 2020).
2. Goyal, M., (2015). Sustainable Micro-Irrigation (pp. 161–170.). Oakville, Canada: Apple Academic Press, Inc.
3. Iqrar, A. K., Muhammad, F., Allah, B., & Muhammad, R. C., (2017). Applied Irrigation Engineering. [ebook] Faisalabad, Pakistan: University of Agriculture Faisalabad Pakistan. Available at: https://www.researchgate.net/publication/325780682_Pressurized_Irrigation_Systems (accessed on 20 August 2020).
4. Jack, K., & Andrew, A. K., (n.d.). Affordable Drip Irrigation for Small Farms in Developing Countries. [ebook] Available at: http://irrigationtoolbox.com/ReferenceDocuments/TechnicalPapers/IA/2003/IA03-0415.pdf (accessed on 20 August 2020).
5. Madhu, M., (2012). ICID: Resources-Irrigation-Pressurized Irrigation. [online] Icid.org. Available at: https://www.icid.org/press_irri.html (accessed on 20 August 2020).
6. Mr. Qazi, S. W. A., (2018). Towards Sustainable Irrigation. [online] Teriin.org. Available at: https://www.teriin.org/blog/towards-sustainable-irrigation (accessed on 20 August 2020).
7. Nejadrezaei, N., Allahyari, M., Sadeghzadeh, M., Michailidis, A., & El Bilali, H., (2018). Factors affecting adoption of pressurized irrigation technology among olive farmers in Northern Iran. Applied Water Science, [online] 8(6). Available at: https://link.springer.com/article/10.1007/s13201-018-0819-2 (accessed on 20 August 2020).
8. Paul, P., Deepak, A., & Bob, N., (1997). A Low-Cost Drip Irrigation System for Small Farmers in Developing Countries (1stedn.). [ebook] Journal of the American Water Resources Association, American Water Resources Association. Available at: http://ftp://ftp.ecn.purdue.edu/vmerwade/class/GDT/uploaded_documents/India/A%20LOW%20COST%20DRIP%20IRRIGATION%20SYSTEM%20FOR%20SMALL%20FARMERS%20IN%20DEVELOPING%20COUNTRIES.pdf (accessed on 20 August 2020).
9. Phocaides, (2007). Handbook on Pressurized Irrigation Techniques (2ndedn.). [ebook] Rome: Food and agriculture organization of the

United Nations. Available at: http://www.fao.org/3/a1336e/a1336e.pdf (accessed on 20 August 2020).

10. Suryavanshi, P., Buttar, G., & Brar, A., (2015). Micro-irrigation for sustainable agriculture: A brief review. Indian Journal of Economics and Development, [online] 11(1), 147. Available at: https://www.researchgate.net/publication/273898393_Micro_irrigation_for_sustainable_agriculture_a_brief_review (accessed on 20 August 2020).

Chapter 8

Innovation and Challenges in Micro-Irrigation

CONTENTS

In the chapter innovation and challenges in micro-irrigation, some of the innovations that have been observed in the micro-irrigation sector is being discussed. The chapter also explains how the micro-irrigation technology comes under the category of an innovative technology.

In addition to this, certain challenges of introducing micro-irrigation in canal irrigation schemes is also discussed in this. Several problems with drip irrigation are also being discussed in this chapter. It also throws some light on issues and challenges of micro-irrigation in improving irrigation water use.

The chapter also explains some micro-irrigation challenge from the perspective of a developing country. Some of the innovative irrigation systems are also being discussed in this chapter. At last, few suggestions for better adoption of micro-irrigation systems is being explained.

8.1. INTRODUCTION

As global concerns about the scarcity of water and food security issues are continuously increasing, the demand for installation of micro-irrigation facility will continue to increase. It is generally seen that Micro-irrigation comes with ample benefits over traditional irrigation methods, incorporating the potential to provide limited supply of water directly to the crop root zone, inclusion of fertigation, reduced weed and pest infestation, and lower operating costs and capital expenditure.

In the current few years, there has been considerable growth witnessed in the acreage under micro-irrigation, majorly due to reduced costs, improvements in emitter and filtration technology, and augmented grower assurance in the technology. Recent research and technological advancement have made micro-irrigation applicable to a more diverse set of applications, water quality conditions and cropping systems. Some critical factors in this is cost and water availability.

Research in bio-filtration and nano techniques, precision irrigation, and soil moisture sensors shows prodigious promise for the progression of micro-irrigation. Nonetheless, various technological challenges will remain in the place, especially for non-orchard or non-row crops, and in regions where water quality is a serious issue. There is significant need of innovations in these areas, as well as a conveyance of the technology to small farmers in water-scarce regions who traditionally surface irrigate.

Hence, there is need of more engineering approaches in order to ensure water productivity is more relevant in economic considerations. Efficient irrigation technologies ensure the real water saving in crop production of semi-arid and arid regions in developing countries. This chapter basically examines the present scenario and innovation in micro-irrigation and appraising the future prospects of micro-irrigation adoption in India.

This paper assesses the current status and technologies of micro-irrigation and evaluating the future prospects of micro-irrigation in several countries especially the developing one.

8.2. MICRO-IRRIGATION: AN INNOVATIVE TECHNOLOGY

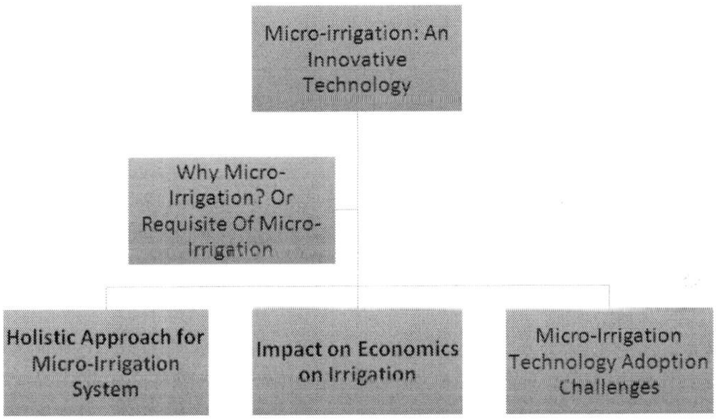

Figure 8.1. Micro-irrigation: An innovative technology.

Micro-irrigation that allow the water to surpass to the root zone of the crops through specially designed equipment is known by the name emitters, which has been widely adopted by many countries for transforming their agriculture. The irrigation sector, mostly because of the tremendously high prevalence of surface irrigation will need to face the difficulty of low WUE in context of the exhausting water resources (Figure 8.1).

The need-based adoption and efficiency of development of micro-irrigation system to address several challenges should be analyzed. As it is experienced in developed and selected developing countries, such as India, micro-irrigation, embracing sprinkler, drip, and mini sprinkler play very significant role in integrated water use system with many socio-economic, agro ecological, and environmental advantages.

It is worth mentioning that micro-irrigation has emerged as a tool that ensures fuller utilization or management of resources that helps in saving fertilizers, water as well as electricity and is one of the interventions in high value agriculture conducive to diversification. In short, it is a multipurpose solution provider-disseminating the advantages entirely, equitably, and evenly in contrast to other irrigation systems. Water productivity is generally perceived as a performance indicator of irrigation in comparison with the old concept of irrigation potential.Water productivity can be explained as the ratio of real agriculture produce and the water actually used. In the case of micro-irrigation, the crop yield is augmented and the water requirement is less when compared with any conventional methods. Hence the productivity of water will be amplified considerably because of rise in numerator and fall in the denominator.

The real water saving in semi-arid and arid regions can be attained in crop production through efficient and effective irrigation technologies. Hence, micro-irrigation has been widely recognized as an innovative technology for sustainable agricultural growth.

8.2.1. Why Micro-Irrigation? Or Requisite of Micro-Irrigation

According to a contemporary survey, it was estimated that 34 countries all over the world will be facing water scarcity situation by the year 2025 signifying that per capita obtainability of fresh water supplies will be less than 100 m3 person per year. A country having availability of renewable water on an annual per capita basis surpassing about 1700 m3 will suffer only infrequent or local water problems.

Below this threshold, countries are more likely to witness regular or periodic water stress. China (1700 m^3) and India (1400 m^3) will come first into this category in the year 2025, while on the other hand, USA will have more than 7000 m^3 person-1 year-1 and will not face the situation of water scarcity.

Rising demand for industrial and urban water supplies across the world throw a serious challenge to irrigated agriculture. The water allocation for agriculture sector will likely to come down to 50% from the existing level of 70%. However, to achieve required food and fiber production with increasing population,

There is need for developing countries to boost the current irrigation potential, especially for those countries that have larger population. The country with higher population such as India is likely to be water stressed in the coming years.

Therefore, coming with the new or advanced technologies for water storage and harvesting for precision water application methods need to be adopted. The critical problem concerned with declining amount of fresh water for irrigation is conveyance losses, dipping the net consumption of irrigation water to 46% only.

However, in order to attain the extra requirement of the irrigation with upgraded technologies for water harvesting, recycling, and storage and excess runoff collection for precision water application by saving the existing amount of irrigation water needs to be adopted. Micro-irrigation will enhance the overall irrigation cover by using the existing available water. Micro-irrigation with Fertigation will likely to increase the production per unit input in these shallow, nutrient deprived, and sloppy soils.

Micro-irrigation is a controlled and co-ordinated water management system where water can be able to flow under pressure through a grid of pipes of varying diameters, the sub-main lines, the main-line, and the lateral lines with adequately placed emitters along the extent of the latter through which water is discharged to the root zone.

It is important to note that the net utilization of irrigation water in drip system is 90% and with the help of sprinkler system, it is about 82%. In reference with the same, micro-irrigation is of high importance with brighter future prospects.

8.2.2. Holistic Approach for Micro-Irrigation System

- Poverty alleviation: it is well known that with the installation of micro-irrigation system, the income of farmers will increase substantially that will help poor farmers to come out of poverty. In addition to this, it will also convert vast barren areas into irrigated areas which are endowed with higher stability and productivity and through the formation of extra employment opportunities, particularly by involving the private sector.
- Horticulture-led diversification of agriculture: Leading to the publicity of high value commodities with incredible prospects for seizing new markets, particularly under the regime of WTO. The system also permits commercialization of protected horticulture with substantial economic returns.
- Enhanced productivity: overall productivity can be enhanced with higher yield by improving the quality of the product and reduced input and labor costs-resulting in intense competitiveness in the liberalized world market. "More crop per drop" must be the driving force.

8.2.3. Impact on Economics on Irrigation

The micro-irrigation technology also has a positive impact on the economics of irrigation as it has an optimistic impact on the pumping hours necessary for irrigating a field. This helps in saving a lot of energy and cost as well as helps in reducing the wear and tear of the pump and escalate its longevity. This is enabled by the reduction in water quantity used for irrigating a field due to the implementation of this technology and this is perceived by both adopters as well as non-adopters.

However, the economics cost might work out but the resource conservation may face challenges as these savings in irrigation would also helpful in expanding the irrigated area thereby increasing the overall quality of water used to earlier levels and also more energy will be needed to irrigate the expanded area. Thus, some advantages of drip irrigation in context to the irrigation economics can be lost substantially, in addition to the better farm economics.

However, ensuring irrigation in this expanded area without drip irrigation would result in significant increase in draught of water that could have significant impact on the environment. Drip irrigation helps in avoiding avoid sort of extremes.

The economics of irrigation is also positively impacted by the technology in many other unplanned ways as it ensures timely and consistent supply of water to the farm; this helps to take away majority of the variability of productivity and farming thereby assuring productivity and yield and revenue to the farmer.The assurance of overall production and yield enables a farmer to enter into long term contracts and make use of great market power as well. It is also seen that the technology has a positive impact on transaction costs of irrigation as it helps in reducing the conflicts and enhances the adaptability in irrigation. However, the real economics of irrigation will get the advantage only once the scale of adoption in time and space is imperative to alter savings into inter sectorial and inter-farm transfers even if only at the household or village level.

8.2.4. Micro-Irrigation Technology Adoption Challenges

It is worth noticing that the adoption of micro-irrigation technology is still a major challenge for developing countries. For instance, Countries such as Indian still continues to be a gamble on monsoon. Only 18% of precipitation is used while the rest is lost through surface runoff, which worsen the existing problem of soil erosion.

Therefore, harvesting the rainwater, monitoring soil erosion with several ways, and efficiently utilizing the conserved/ harvested water, needs to be united in a way to validate the significant investments on high cost irrigation projects. In the non-existence of large-scale adoption of well-organized modern irrigation technologies such as sprinkler and drip irrigation, the existing irrigation efficiency in the country is only 30 to 40%.

Past efforts in endorsing micro-irrigation technologies would be turn as rewarding and they did bring vertical growth in their adoption. However, their adoption has stayed restricted with slow progress, because of the following reasons:

- The preliminary cost for establishment of micro-irrigation system is high, hence usually out of the reach of poor farmers deprived of resources.
- Micro-irrigation technology was not united with farm irrigation management systems, as it is often observed in the isolation.
- Proper loan facility to the farmers, well skilled and experienced human resources, appropriate infrastructure and technical know-how for training of farmers were lacking.
- Micro-irrigation is often considered as technology intensive, hence its recognition by farmers requires much persuasion.
- There was a lack of information on spatial and temporal variation in soil moisture, the ideal fraction of soil to be drenched, crop-specific, and location specific irrigation and Fertigation scheduling and lack of economic water-soluble fertilizers availability and other agro chemicals.

8.3. CHALLENGES OF INTRODUCING MICRO-IRRIGATION IN CANAL IRRIGATION SCHEMES

India utilizes the most significant measure of water (557 BCM) for irrigation on the planet. According to the Ministry of Water Resources, the water required for irrigation will increase to 611BCM in 2025, and 807 BCM by 2050.

Groundwater has around 60% of the water irrigation system uses and the portion of drain flooded area has been constantly declining. Because of expanding freshwater request from household and mechanical areas, the portion of water used in the irrigation system will diminish further.

To adapt up to the rising water request from horticulture, the utilization of dribble and sprinkler water system has been consistently expanding in India. The trickle water system was presented in India on business scale in the year 1985 when the territory inclusion was negligible 1500 ha.

During the decade 1990–2000, the dribble watered zone drastically expanded multiple times for example from 0.035 million ha to 0.351 million ha; while in the ensuing decade (2000–2010) it developed by in excess of multiple times for example 1.9 million ha in 2010.

At present, it is evaluated that 3.0 million ha are trickle watered everywhere throughout the nation putting India as the most elevated smaller scale flooded nation on the planet. Both, administrative, and state Governments are giving liberal impetuses as appropriation towards establishment of smaller scale and sprinkler water system.

At present, dominant part of the miniaturized scale water system relies upon groundwater and just some irrelevant region is on channel water. Execution of the greater part of the general population managed irrigation plots in India is poor that is prompting to low water use productivity.

Farmers, because of untrustworthy waterway water gracefully, as a rule apply more water to their yields prompting wastage of water and decrease in arranged territory for irrigation. In addition, the expense of development of dispersion organize, on-ranch structures and their resulting maintenance is getting progressively restrictive.

A couple of states like Gujarat, Himachal Pradesh, Karnataka, and Rajasthan have received miniaturized scale irrigation in the waterway orders. As of late, Maharashtra State has taken a choice to receive miniaturized scale irrigation in trench order zone of select irrigation extends on pilot premise.

The significant target behind this activity is to decrease the on-ranch water misfortunes and utilize the spared water to denied territories, especially tail comes to in an irrigation plot. According to the notice brought out by the Water Resources Regulatory Authority of the State, it is compulsory to receive small scale or sprinkler irrigation for money and high-water requiring crops like Sugarcane and Banana.

The extent of such harvests in the irrigation plans of the State regularly extend from 10 to 20% of the watered region. This impulse has set off a discussion as there are numerous difficulties to get it going. This involve changing over and working the common trench system in half and half mode, both as surface just as smaller scale irrigation systems.

Some of the major issues that are required to be addressed:

1. Guaranteed water accessibility at a shorter time period to 3 days;
2. Stockpiling at farm level;
3. Continuous accessibility of electricity;
4. Money related help as sponsorship towards capital expense of the small-scale irrigation system; and
5.` Implementation at the ground level.

The key issue which should be addressed is whether there would be 'genuine water saving' and can the spared water, if any future made accessible for different farmers or different employments. It is considered to present the smaller scale and sprinkler irrigation in three significant trench irrigation plans including one lift plot.

In light of the experience from these pilot plans, it will be upscale in the other irrigation plans of the State. This chapter will give a short situation of smaller scale irrigation improvement in India and a trivial conversation on the different potential limitations and route forward in execution of miniaturized scale irrigation in the current waterway order territories.

8.4. PROBLEMS WITH DRIP IRRIGATION

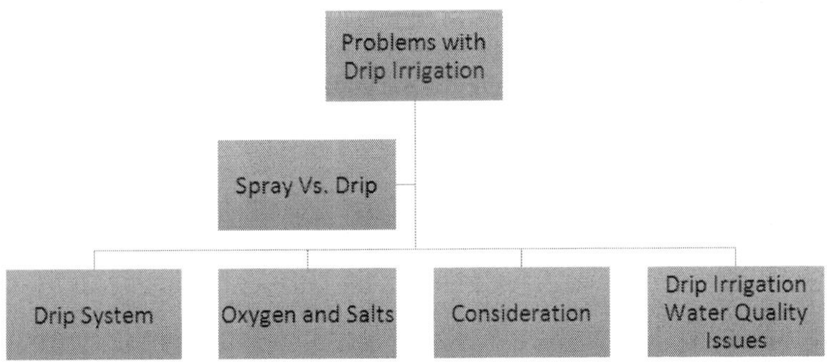

Figure 8.2. Problem with drip irrigation.

Underground irrigation systems were widely used in the past few years and was once the only available option for agriculture activities-until drought, water scarcity, and comprehensive understanding about how plant diseases work invigorated the advance of more effective watering methods (Figure 8.2).

Of course, with the passage of time, innovative manufacturers introduced some advanced irrigation system, declaring its superiority to soaker hoses and sprinklers alike. These new low-pressure systems ensure more effective and efficient use of water but comes with their own set of problems. Some of these problems are discussed as follows:

8.4.1. Spray Vs. Drip

Spray irrigation, which is formed for connection in turfgrass, is an unproductive way to transport water to agriculture land.

A major proportion of water unavoidably evaporates as it arcs through the air, even when it is properly controlled.

This spread the water through the air which result in fall of water to more unnecessary things rather than directly poured into roots of crops. Finally, water splashes in the dirt, taking with them soil-borne bacteria and fungi, transporting the pathogens to the soil of crops.

Soaker hoses spread the water directly on the ground, but they may supply more waters to areas where there is no need of water, excessive supply if water may result in wash soil away from plants and roots.

Drip technology steps down water pressure, plummeting the 60 to 180 gallons of water used by a sprinkler to 1 to 4 gallons each hour by a drip emitter. Regrettably, a drip system demands a new set of pressure emitters, converters, and 1/2-inch line for effective operation.

8.4.2. Drip System

A drip system gives water where it's required, but since of its low weight, cautious guideline must give enough water to support sound root advancement. To answer requirements for more profound watering and overhead watering, makers change it up of weight controllers; for ground covers, they make miniaturized scale sprinklers that utilization from 5 to 45 gallons for every hour.

In any case, drip systems must run for additional hours than a conventional system to accomplish profound watering, and managing a system where results are not evident requires some investment and consideration, especially when hard water obstructs pressure valves.

8.4.3. Oxygen and Salts

A similar sprinkler splashing that builds moistness and moves soil-borne pathogens to plant surfaces similarly drives oxygen into the dirt and powers salts from water and composts to permeate descending, conveying the previous and disseminating the last mentioned.

Drip irrigation's delicate nature does not one or the other. Visit development drives oxygen into the dirt and profound watering with a hose flushes the dried-up white ring, much like the white outside that structures on houseplant pots and soil, through the dirt. Small scale sprayers help with soil oxygen, but since just a constrained territory is inundated, the salt rings remain.

8.4.4. Consideration

The complete water utilized in a drip system might be not exactly a conventional irrigation system, yet parts can be costly and may not generally be good. Drip system units that associate directly to outside plugs may introduce an alignment challenge to the nonprofessional.

Technology used in this is consistently involving, often resulting in causing expensive equipment to become obsolete from year to year. There are few manufacturers that tried to solve the compatibility problem by designing emitter hoses, pressure valves, and sprinklers that are adequate in standard irrigation systems, providing the existing system is one of theirs.

8.4.5. Drip Irrigation Water Quality Issues

Figure 8.3. Drip irrigation water quality issues.

Source: Image by Flickr.

Water quality can negatively affect the irrigation system's working due to stopping of sprinklers and emitters. Issues can be brought about by inorganic solids, natural solids and broke down solids (calcium, iron, manganese) (Figure 8.3).

Potential issues can be limited by testing your water so you know about potential issues. Filtration is the essential prerequisite for all drip systems to evacuate any suspended solids. This can be cultivated with one or various filtration strategies relying upon the water quality issue.

Sometimes, compound water treatment may likewise be required to keep up uniform water application. Chlorine is the treatment concoction used to control microbial development, break down natural materials, and oxidize dissolvable minerals.

Potassium permanganate can likewise be utilized to oxidize iron. Oxidation will make minerals hasten out of arrangement so filtration must follow treatment. Corrosive medicines can be utilized to bring down water pH to keep up solvency of minerals or to break down mineral built up around producers.

Hardware for water treatment is like chemigation gear. Filtration ought to consistently follow any water treatment to expel any hastens.

8.5. ISSUES AND CHALLENGES OF MICRO-IRRIGATION IN IMPROVING IRRIGATION WATER USE

In spite of the plethora benefits of modern irrigation methods such as sub-surface or surface drip and sprinkler irrigation systems, they are not extensively executed by farmers on large scale because of significant initial capital investment, uncertain water supply for the system, scattered, and marginal agricultural landholdings, nonexistence of subsidy at the right time or the delayed release of funds for installation of micro-irrigation system that have been approved already, and no easy financing options for farmers.

Another major challenge for such poor adoption of technologies is because of farmers' inclination towards more traditional methods of irrigation as they do not possess proper knowledge about the benefits and opportunities provided by the micro-irrigation system.

It is the duty of the concerned department to make them aware about the same so that they can take the decisions that benefits them in the long term. In addition, because of such recurrent changes in schemes/ components concerned with National Mission on Micro-Irrigation (NNMI), various schemes introduced by the government of developing countries like India from time often lack proper support and focus to dedicated work towards this area.

Some of the major issues that are faced by government especially government of developing countries in implementation and growth of micro-irrigations system are discussed as follows:

8.5.1. Tedious Financing Machinery to Beneficiaries

It is often seen that providing financing options to beneficiaries conducive to lot of hindrances, as a result of which they unremittingly face problems in securing financing options. There is an existent difficulty with the inaccessibility of funds in some states, resulting in more delay in disbursement of fund from the state government to the concerned state or village in which farmers require to implement micro-irrigation system.

It generally causes obstruction in promoting and initiating micro-irrigation system schemes at the state level. This as a result make it more difficult to distribute subsidies to needed farmers, that usually ranges from 120 to 150 days, whereas need to be 75 days.

This is the most prevalent scenario in developing countries like India where there half of the population is associated with agriculture sector. There is an utmost need to find ways to guarantee availability of funds on time and for simpler funding norms for farmers on priority

8.5.2. Insecurity and Periodic Variations in Irrigation Scheme Strategies

Owing to the lack of long-term and effective strategies, most of the schemes employed by countries are operational only for a restricted period of time at the state level. Due to this, for instance, Federation of Indian Chambers of Commerce and Industry (FICCI) report has revealed that on an average basis, schemes are operative only for 5 months in a year to the farmers.

As a result of this, farmers could not avail the benefits of the scheme at the time when demand of water is high. In addition, this may force the farmers to miss the cropping season, unable to extract the maximum benefits from farming season, leading to their inability to gauge the real benefits of the micro-irrigation system.

8.5.3. Scheme Execution and Functioning Processes

It is worth noticing that majority of the micro-irrigation schemes are inefficient and ineffective in relation to their functioning and execution, especially in terms of the overall time taken and length of the process of scheme execution.

In some states of developing countries, the launch process and scheme implementation occur during the months of August and September, because of this, there is not much needed supply of water for irrigation and timely irrigation to the field. By the time a scheme that particularly aimed at providing needed water to crop came into functioning, the real season already over.

Therefore, beneficiaries are not able to extract the maximum opportunities and benefits from the existing micro-irrigation technology. Thus, there is an utmost need for timely completion of the process (foremost before monsoon season) for better coordination with the farmers' demands and interest.

Lack of a proper monitoring system makes the task of functioning and managing scheme execution extremely challenging. This has a significant impact on the efficiency of implementing micro-irrigation system.

8.5.4. Lack of Motivation on Micro-Irrigation

It is important to note that a diluted focus on the micro-irrigation scheme is still in existence, with the government of respected countries ensuring allocation of funds for micro-irrigation system since 2014–15 as a component of government schemes and not through the committed mission such as the erstwhile National Mission on Micro-Irrigation (NMMI).

This is majorly followed to a greater extent at the state level as there are only limited states having a dedicated team to encourage micro-irrigation, while large number of others do not make that much efforts that needed, resulting in deceleration in the growth of MIS.

8.6. THE MICRO-IRRIGATION CHALLENGE FROM THE PERSPECTIVE OF A DEVELOPING COUNTRY

Figure 8.4. Huge set up cost of micro-irrigation, a challenge for developing countries.

Source: Image by business line.

An example of developing countries like India

It is generally seen that in developing countries like India government are spending huge amount in providing micro-irrigation facility to agriculture sector. As agriculture sector is their significant source of income, there economy much or less rely on the agriculture sector. In fact, the people of developing countries are highly depended on their agriculture sector for their food supply (Figure 8.4).

For instance, In India, government had allocated over Rs 5,000 crore was allocated to micro-irrigation, up by 276% from the UPA-II regime (2009–14)-long overdue in the backdrop of unprecedented crop failure in the past couple of years and burgeoning agrarian unrest with dire socio-economic ramifications.This move receive acceptance from almost all experts of the country. Across the world, most nations are now increasingly leaning to micro-irrigation-based strategies as it provides substantial water-energy savings, resulting in significant economic gains on input.

According to a study conducted by Grant Thornton India L.L.P. in the year 2016, in collaboration with the Indian Agricultural Institute (IAI) and FICCI estimated that micro-irrigation can lessen present energy demand in the agriculture sector by almost 30%, and the overall expenditure incurred will be reduced by almost 28%, besides increasing opportunities of introducing newer crops.

But one of the most important question that arise in this area is that are the aspirations grounded in reality?

As a first reality check, the Thornton report suggests that with the current speed at which micro-irrigation facility is implementing in agriculture facility of India,–0.5 million hectare/annum-it will take almost another 100 years to provide 'nationwide' micro-irrigation support.

Thus. It can be concluded that in developing countries, it is not an easy task to provide micro-irrigation to agriculture sector on such a large scale, as it requires huge investment and government is in shortage of funds.

According to a recent report of the International Energy Agency (IEA), it was projected that India will face intense power shortages in the next few decades, attributable largely to water shortage. Water is used as a coolant in coal-fired power plants.

With a minimum 19 states out of 29 states comes under acute water shortage (affecting livelihoods of 330 million people), carrying agriculture activities in these areas is really a challenging task, pitting in turn, nation's power supply framework at the very core.

In 2016, there was nationwide loss of more than 7 billion units of power, as a recent Greenpeace report estimates. Number of big power plants were forced to shut down, triggering huge financial losses and widespread power supply crises.

According to a FICCI report estimated that in 2013, power shortages cost 0.4% of (USD 68 billion). In the year 2012, there was a loss of about 12,000 MW of power. With hydroelectric working with its full capacity to the limits due to reduced stream-flow, the imbalance between demand and supply trigger the increase in price owing to constant operation of irrigation pumps and air-conditioners.

Dire electricity crunch-characterized by rampant pilferage (illegal conversions/ diversions), aberrant voltage fluctuations, inadequate metering facility-are already significant hitches to irrigation pumping in India, and by the similar way, could be greatest hindrances to the ambition for enhancing micro-irrigation in the country.

In most part of the countries, farmers are mostly keeping their pumps turned on for 24*7. There are various instances known where the system has been automated so that pumping resumes its operation immediately after electricity arrives. These kinds of actions often resulted in faster depletion both groundwater resources and electricity.

As there is drop in water level because of withdrawing of water at such raid rate, it necessitates the need for irrigation pumps run at almost all the time to tap deeper reserves, thus straining both water and power sector to the limits.

The various eco-environmental challenges arising from increasing dimensions of such water-energy connection in the irrigation sector assumes added importance with extreme drought alert in numerous states accompanied with intensifying land desertification-degradation interfering with daily livelihood opportunities of millions.

But apart from such technological challenges, there are various policies associated with getting out from this difficult situation.

For instance, micro-irrigation schemes have consistently changing with context to the current situation and needs (or been 'subsumed' under multiple schemes) over time-beginning with the Centrally Sponsored Scheme on Use of Plastic in Agriculture (1992) and accompanied by Accelerated Irrigation Benefit Program (AIBP, 1996); National Horticulture Mission (2005); Centrally Sponsored Scheme on Micro-Irrigation (2006); Rashtriya Krishi Vikas Yojana (2007), National Mission on Micro-Irrigation (NMMI, 2010); National Mission on Sustainable Agriculture (2014) and finally to its present form, Pradhan Manti Krishi Sinchayee Yojana (PMKSY) in 2015.

One important question that come in this way is whether amidst merging and re-merging through the years, has cause significant impact of micro-irrigation as a stand-alone programme been 'diluted'?

Various school of thought came to conclusion that such change may have resulted in achieving the needed level of capacity building-a dedicated team of trained/experienced professionals particularly assigned to address micro-irrigation linked demands (research and development), and/or systems to back it up by necessary IT support.

The latter believed to be of high importance as it is concerned with ensuring real-time monitoring of transactions and geo-tagging-two important elements in successful implementation of micro-irrigation projects.

Other major challenges include in this are hindrances include,

1. Delays in providing subsidies to those farmers who want to install irrigation system at their land-primarily due to approving installation of equipment when the available funds to deliver

subsidy aren't yet available, that often resulted in a shortage of funds, and in turn, cause uncertainty for the suppliers of micro-irrigation systems and delays in subsidy payments.

2. Lack of extensive private financing programs for micro-irrigation projects (NITI Aayog have allowed private sector proposal in 2016); and
3. Lot of paper work for farmers if they want to get finance for irrigation facility, in addition they are required to submit collateral if they want loan, making it utmost difficult for farmers to get loan.

Another major challenge is about the economic viability of these developing countries. Yet in many developing countries like India, micro-irrigation peripherals are quite expensive for the smallholding communities.

In addition to this, there is no proper support for maintenance (for instance, rodent attack on pore-clogging, piping) and operation.

Another challenge that necessitate the need for serious policy intervention is the 'mismatch' between seasonal cropping patterns and timing of micro-irrigation supports access to the farmers.

It is well known that Micro-irrigation schemes are yet not operational year-round (on an average available for just five months in a year), which means farmers are more at a risk to miss the opportunity of the cropping season, not able to extract the maximum benefits and thus, gradually become disinterested.

Overall, multidimensional challenges, varying from negative corollaries between the energy and water sectors to lags in policy framework, deteriorate future opportunities of expansion of micro-irrigation network. It is largely depending upon can effectively turn these challenges into opportunities and come up with innovative and feasible solutions in the next few years.

8.7. MOST INNOVATIVE IRRIGATION SYSTEMS

Figure 8.5. Looking for innovative ways for irrigation.

Source: Image by USDA.

The irrigation system comprises a progression of segments, some inventive, in spite of the fact that not really the irrigation system should comprise every one of them, since the arrangement of parts will rely upon whether it is surface irrigation by drip or by spray (Figure 8.5).

The issue of sparing water is substantially more sensitive than it appears, in this way, all that can be done to add to a mindful utilization of this asset, and it should be exploited. The most creative irrigation systems have been completely devoted to the job and not simply to make things simpler to deal with the harvests and manors.

8.7.1. Irrigation Programmers

Irrigation developers are supervisory groups that permits to advance the time and the zone of irrigation in our harvests, permitting more prominent water reserve funds and improving our solace. Moreover, we can mechanize the recurrence and length of irrigation.

As to type, there are simple and computerized irrigation software engineers. Analogs are simpler to utilize and are less expensive, yet advanced ones for the most part have more capacities.

At last, it ought to be noticed that there are developers with at least two courses that permits to plan distinctive irrigation programs for various regions, which is ideal for huge yields or zones with various water needs.

8.7.2. Solar Irrigation

Figure 8.6. Solar irrigation.

Source: Image by Flickr.

Solar watering, otherwise called condensate compressor, is a foundation with two plastic containers that depends on vanishing and buildup. In the inside, the nursery impact is delivered, as the solar vitality impacts and creates said dissipation (Figure 8.6).

At the point when the inside air gets immersed with dampness, water consolidates on the dividers of the container and structures drops that fall by gravity to the ground. In the inside of this irrigation system, a little scope water cycle is delivered, giving water to the dirt ceaselessly.

This irrigation system is extremely easy to make and is inventive in irrigation since it permits more noteworthy reserve funds, which comes to lessen multiple times the measure of water utilized in customary irrigation systems.

8.7.3. Sprinkler Irrigation System

Sprinkler irrigation systems gather water from a well or directly from a stream or repository. The water saving is more as compared to different systems. It tends to be utilized on bumpy landscape, offers full inclusion, and conveys water easily. It can be recognized by the following systems (Figure 8.7):

Figure 8.7. Sprinkle irrigation to spread water to a greater extent.

Source: Image by city of college station.

Traditional Systems: First irrigation systems created in sprinkler irrigation. They are separated into:

Fixed Systems: the sprinklers are put in the built-up outline, and the funnel system can be covered or shallow, leaving as a jutting part and with the suitable tallness the stem where the sprinkler will be fused.

Systems Semi Fixed: they are basically systems that are moving starting with one zone then onto the next physically or automated by fast dismantling of the system.

Auto mechanized systems: they are car systems that have introduced electric engines or water driven systems that permit their development along the irrigation surface. Inside these systems are the turning systems of irrigation, horizontal uprooting systems and other watering machines.

8.7.4. Irrigation System by Micro-Sprinkler

The miniaturized scale sprinkler irrigation system is a variation of sprinkler irrigation that has a littler range and appropriates lower beads, permitting more water investment funds. For these attributes it is a standout amongst other irrigation systems for agricultural yields, blossoms, nurseries, or nurseries, etc.

8.7.5. Drip Irrigation System

The drip irrigation system is normally utilized in dry zones since it permits a superior use and investment funds of both water and composts. The water through it is invaded towards the underlying foundations of the plants by methods for a system of drippers. Development in the most significant irrigation in the agribusiness area is considered since the sprinklers were imagined in 1930.

8.7.6. Underground Drainage in Agricultural Land

Figure 8.8. Underground drainage system.

Source: Image by Pikist.

The goal of underground waste is to bring down the profundity of the water table, which is the upper layer of a spring.

It is normally utilized in agrarian fields that are excessively damp or whose groundwater levels are unreasonably high for harvests to develop (Figure 8.8).

8.7.7. Multi-Door Irrigation System

The multi door irrigation system directs the water and disperses it over the land assisting with streamlining the utilization of this asset. Of all the irrigation systems referenced, this is the least known and furthermore the least utilized.

8.7.8. Airdrop

It is an extremely inventive irrigation system dependent on the buildup of water, diverting the air through a turbine to a water supply that is situated in the dirt. In transit, the air cools and consolidates before arriving at the tank. When the water is in, it is siphoned to the drip irrigation system giving water to the plant.

It is a development in irrigation since they normally fuse both regular and man-made materials. All around, it is based, other than the plants, on the pergolas, which are spaces between gardens; they can connect structures and make coverings on the entryways of structures, on porches or pools.

8.7.9. Increase in Efficiency

The insufficient utilization of water not just wastes the asset that could be utilized for different purposes and to help evade natural effects, downstream, yet in addition purposes weakening, through immersion, salinization, and filtering, and decreases the efficiency of harvests.

The improvement of water use, in this way, ought to be the primary worry of any irrigation system.

There are enormous territories of land under irrigation that have quit creating because of soil decay.

It might be helpful and, obviously, gainful to the earth, to put resources into the reclamation of these terrains, as opposed to expand the territory of low irrigation.

8.8. SUGGESTIONS FOR BETTER ADOPTION OF MICRO-IRRIGATION SYSTEMS

It is important to note that the large-scale adoption of micro-irrigation technologies is of utmost important and should be backed by government of the respected country in a way to extract the maximum benefits offered by these technologies.

- Micro-irrigation should never be considered as an end in itself. It is required to be promoted in a holistic manner having end-to-end approach with linkages for marketing and incorporation of agronomic practices.
- Technologies of micro-irrigation technologies should be considered as total plant support systems rather than just viewing them as water conservation measures, as these technologies help in enhancing the combined and individual efficiency of all other agricultural inputs such as energy, fertilizers, plant protection measures, etc.
- It is widely believed that Major & Medium irrigation projects have generated islands of prosperity amid existing poverty.

It is because of the fact that, the upstream population and the population beyond the tail end of the canal have been divested of the plan advantages.

The depressed lot has also been lost in the meanwhile even what they once received as their ordinary portion in the water flowing through the river. Farmers that are living on upstream and beyond tail-end have left with no option but to restrict to cost intensive lift irrigation schemes.

Even here, inept water uses through conventional methods conducive to energy costs and prohibitive water rendering the scheme uneconomical. In case, government in various countries make micro-irrigation technology obligatory for all commercial crops under both lift and canal irrigation system, it will definitely provide large significant benefits to large number of farmers with same volume of water. This will obviously lead to greater equity.

- In developing countries, it is often seen that groundwater is the only water source being used for drip method of irrigation. In majority of the countries, water from surface sources (reservoirs, dams, etc.) is generally being used for drip method of irrigation while for low income countries, it is not the case.

Since the overall efficiency of water under surface sources is very low resulting in heavy losses through distribution and conveyance, farmers in these countries must be encouraged to use water from surface sources for drip method of irrigation. It could be possible only by assigning certain proportion of water from each irrigation project only for the use by micro-irrigation.

- Water Users Associations (WUAs) are being endorsed in a big way with the support of external and internal agencies in recent years. Since one primary objectives of the WUAs is to enhance the overall efficiency of water, some arrangements need to be made to introduce the micro-irrigation through WUAs.

As there is always shortage of electrical energy in the developing countries, its conservation in the agricultural sector backed by the micro-irrigation should receive due consideration even in designing future policy of power.

- Micro-irrigation often inspires commercial agriculture through diversification of crops and persuades the farmers to shift from conventional cropping methods to high value commercial crops. It also enhances quality of the produce because of uniform and timely nutrient application.

- Capacity building of the manufacturers play very important role as it allows to develop quality and advance products and their application. There should be consistent engagement between the research organizations and micro-irrigation industry for updating the skills and knowledge.

8.9. CONCLUSION

In the end, it is concluded that micro-irrigation is one of the critical factors for the overall success of agriculture sector. The role of micro-irrigation is more important in developing countries as their income is heavily dependent on this sector.

The recent innovation in the micro-irrigation has made certain tasks of agriculture activity easy such as solving problem of water supply to agriculture land. However, all this innovation comes along with various challenges that need to be faced by the famers whether it is from acquiring the technology to continuous maintenance of technology. Although, the benefits generated from this technology outweighs its challenges. So, it can be said that micro-irrigation is the much-needed fuel which is essential for the smooth performance of agricultural activity.

REFERENCES

1. (2020). Problems and Prospects of Irrigation. [ebook] Available at: https://sg.inflibnet.ac.in/bitstream/10603/95523/16/16_chapter%208.pdf (accessed on 20 August 2020).
2. Academia, (2018). Wproblems and prospects of Adoption of Drip Irrigation System in Tamilnadu. [online] Available at: https://www.academia.edu/37389687/PROBLEMS_AND_PROSPECTS_OF_ADOPTION_OF_DRIP_IRRIGATION_SYSTEM_IN_TAMIL_NADU (accessed on 20 August 2020).
3. Chaudhuri, S., (2018). The Micro-Irrigation Challenge. [online] The Statesman. Available at: https://www.thestatesman.com/opinion/micro-irrigation-challenge-1502599259.html (accessed on 20 August 2020).
4. Gardening Know How, (2020). Problems with Drip Irrigation-Drip Irrigation Tips for Gardeners. [online] Available at: https://www.gardeningknowhow.com/garden-how-to/watering/drip-irrigation-problems.htm (accessed on 20 August 2020).
5. Gestiriego, (2019). Most Innovative Irrigation Systems-Gestiriego. [online] Available at: https://www.gestiriego.com/en/most-innovative-irrigation-systems/ (accessed on 20 August 2020).
6. Goyal, M., (2020). Innovation and Challenges in Micro-Irrigation. [online] ResearchGate. Available at: https://www.researchgate.net/publication/289460778_Innovations_Challenges_in_MIcro_Irrigation (accessed on 20 August 2020).
7. Green Growth Knowledge Platform, (2016). Micro-Irrigation: Trickling-Down Inclusive Green Growth. [online] Available at: https://www.greengrowthknowledge.org/blog/micro-irrigation-trickling-down-inclusive-green-growth (accessed on 20 August 2020).
8. Grounded Landscaping, (2020). Top 10 Irrigation System Problems and Solutions-DIY-Grounded Landscaping. [online] Available at: https://www.groundedlandscaping.co.za/top-10-irrigation-system-problems-solutions/ (accessed on 20 August 2020).
9. Inocencio, A., Merrey, D., & Sally, H., (2020). Innovative Approaches to Agricultural Water Use for Improving Food Security in Sub-Saharan Africa. [ebook] Available at: https://www.joinforwater.ngo/sites/default/files/library_assets/W_LAN_E10_innovative_approaches.pdf (accessed on 20 August 2020).

10. Madramootoo, C., & Morrison, J., (2013). Advances and challenges with micro-irrigation. Irrigation and Drainage, 62(3), 255–261.
11. NoCamels - Israeli Innovation News, (2018). From Drip Irrigation to Digital Farming, Netafim Delivers Innovative Agri-Tech Solutions Worldwide | Environment News. [online] Available at: https://nocamels.com/2020/01/drip-irrigation-digital-farming-netafim-innovative-agriculture-tech/ (accessed on 20 August 2020).
12. Reynolds, L., (2018). Problems with Drip Irrigation. [online] Home Guides | SF Gate. Available at: https://homeguides.sfgate.com/problems-drip-irrigation-77349.html (accessed on 20 August 2020).
13. Sharma, S., &Suhirid, M., (2020). Micro-Irrigation an Innovative Technology- its Importance, Challenges and Present Scenario in India. [online] Icid.org. Available at: https://www.icid.org/9imic2019_papers/9th_imic_st1_paper%20(13).pdf (accessed on 20 August 2020).
14. Understanding Crop Irrigation, (2018). Drip Irrigation Water Quality Issues. [online] Available at: https://fyi.extension.wisc.edu/cropirrigation/drip-irrigation-water-quality-issues/ (accessed on 20 August 2020).
15. Vaibhav, R., & Makwana, A. D., (2018). Impact Analysis of Micro-Irrigation System (MIS) On Yield, Water, Fertilizer Saving and Farmer's Economy. [ebook] Available at: https://www.ijcmas.com/7-10-2018/Ram%20Vaibhav%20and%20D.%20Makwana%20Ajay.pdf (accessed on 20 August 2020).

INDEX